Sasquatch
in Alberta

Thomas Steenburg

CRYPTO
editions

ISBN-13: 978-0-88839-408-8 [trade edition softcover]
Copyright © 2018 Thomas Steenburg

Cataloging in Publication Data

Library and Archives Canada Cataloguing in Publication

Steenburg, Thomas N. (Thomas Nelson), author
Sasquatch in Alberta / Thomas Steenburg. -- Trade edition.

Originally published under title: The Sasquatch in Alberta.
Calgary: Western Publishers, 1990.
ISBN 978-0-88839-408-8 (softcover)

1. Sasquatch--Alberta. I. Title.

QL89.2.S2S76 2018   001.944   C2018-904609-0

Illustrations and photographs are copyrighted by the artist.

Printed in the USA

PRODUCTION: M. Lamont
FRONT COVER: "Encounter" by Tim Heimdal, Grande Prairie, Alberta

Crypto Editions is an imprint of Hancock House Publishers

Published simultaneously in Canada and the United States by

HANCOCK HOUSE PUBLISHERS LTD.
19313 Zero Avenue, Surrey, BC Canada V3Z 9R9
(604) 538-1114   Fax (604) 538-2262

HANCOCK HOUSE PUBLISHERS
#104-4550 Birch Bay-Lynden Rd, Blaine, WA U.S.A. 98230-9436
(800) 938-1114   Fax (800) 983-2262

www.hancockhouse.com   sales@hancockhouse.com

ABOUT THE AUTHOR

Born in the province of Ontario, 1961. Thomas Steenburg has been actively interested in the subject of the Sasquatch since the age of five. In 1980 he enlisted in the Canadian Army with the Princess Patricia's Light Infantry, First Battalion stationed in Calgary, Alberta, and since 1979 has actively gone into the rugged outdoors in search of the elusive creature. Having interviewed many eye witnesses and investigated many reported sightings it has made him all the more determined to pursue and answer as to whether the existence of the Sasquatch is real or simply folklore.

It is his hope that someday hard physical evidence will be found to once and for all solve this mystery for all of us.

FOREWORD
Vladimir Markotic

I have been interested in the Sasquatch problem for a long time. I have a degree in Anthropology and an interest in primates comes natural to me.

In my opinion, there is nothing impossible about the Sasquatch. The descriptions fit well with studies of the anatomies of apes and men. The only extraordinary thing about them is that they are so tall. But their tallness corresponds well with the supposed casts of their feet which are larger than those of apes and humans.

I do not believe that a Sasquatch could be ten feet tall, and the height of fourteen feet is out of the question. Judging the heights of unknown objects and beings is a very tricky thing. The most amazing case of wrong perception involved the Pygmies in Central Africa, who live in the forest. When a Pygmy was brought out of his natural habitat into the treeless plains and saw buffalo grazing several miles away he asked "What insects are those?" because he had no sense of distances. Living in the jungle Pygmies are used to distances which extend only to the next trees. (Colin M. Turnbull: The Forest People, p. 263.)

Even more difficult is to judge objects when they are not on the same level as the viewer but are higher or lower. We experience this when we drive to and through the rocky Mountains. We see trees that we know are tall but look like small toys. Slavomir Rawitz was a Polish officer who escaped from a Siberian prison camp during the last war and crossed the Himalayas on the way to India. There he saw two Yeti, (a creature somehow related to the Sasquatch). He wrote, "I set myself to estimate their height on the basis of my military training for artillery observation". Their heights according to him, should not have been much less than eight feet. (Slavomir Rawitz: The Long Walk, P. 228)

I mention these facts to explain why there are some discrepancies or variations when people described the size of Sasquatches. These variations are due to different perceptions, different cultural backgrounds of the viewers, whether they observe in the daylight or at night, etc. But on the whole the Sasquatch as he emerges from all these reports is a pretty uniform being. He is between six and eight feet tall, hairy except on the face and palms, has even length of teeth, human-like feet, where the first toe is not separated from the other parts of the foot as among the apes. He walks straight on two feet only. This last characteristic (with the anatomical similarities of his feet to the human) and the evenness of his teeth makes it possible to classify him even more precisely as a hominid.

The uniformity of anatomical details about the Sasquatch make me think it is not impossible for Sasquatches to exist. As a matter of fact, I would be surprised if they didn't.

They appear as straight biological entities and not as some strange misunderstood phenomena. Thus, the possibilities of the Sasquatch can be treated statistically and scientifically. Tom Steenburg does so by giving at least two sides of each case he presents. He also investigates each case himself and talks to as many witnesses as he can find. He checks and re-checks and investigates the story from all possible angles.

This book may bring surprises to those of you who do not know there are many stories about the Sasquatch in Alberta and not just in the mountainous regions of western British Columbia. Tom Steenburg is to be commended to have put these stories from Alberta together and I recommend this book to anyone interested in the Sasquatch.

Vladmir Markotic. Photo taken by author in 1984.

iii

INTRODUCTION

Was it merely an Indian Legend told to early explorers, a story characteristic of the native's mythical culture? Or, were the stories of giant hairy man-like apes actual reports of an animal that has managed to mystify its researchers and elude Western civilization for nearly three hundred years?

While the European pioneer travelled across the continent of North America he was entertained by its colorful natives and their flashy heroic tales of this elusive man-like beast. The names for this animal were as varied as the tribes of the North American Indians, themselves. Sasquatch (English pronunciation) was only one such name.

The same cultural differences that had entertained the settler, however, were to become part of the settlers' intolerance of the Indian ways. When the natives refused to give up their stories, religious beliefs and land and join the white man's "more civilized approach" to life their cultural structure was first discredited and then destroyed. The Indian was eventually cast away onto "reserved land" along with his traditions, lifestyles and legends.

Long after early explorers had laid claim to and secured their lands and cities, however, the descendants of the settlers began moving back to unpopulated territories to enjoy the beauty of the wild mountains and attempt to preserve what was left of the forests. It was then that the Legend of the Sasquatch was reborn.

Campers, hunters, adventurists were returning home with incredible reports of one or more giant hair-covered man-like beasts, huge man-like footprints and strange unidentified animal calls. Only this time, the "civilized man" would not hear; no such animal had ever been seen publicly or found by the scientific community. The Sasquatch became a joke, a myth, a money making event, a mere speculation to all but a select few men whose determination to prove its existence would not be quenched.

Several years ago, two such men grabbed the attention of the world with a film they had made of the elusive creature. Try as it might, the scientific community could find no evidence to excuse the film as a hoax or a set-up. As legitimate explorers, researchers and a few scientists joined in on the hunt to find the Sasquatch, the race for evidence was on; to prove the existence of and settle the Sasquatch mystery once and for all.

John Green, a Canadian widely renowned for his research into this subject proposed the theory that the Sasquatch is an ape, a great ape residing within but not exclusive to the North American continent. Professor Grover Krantz, an anthropologist from Washington State University has, from his studies, come to the very same conclusion. From his research into fossil records, Professor Krantz has surmised that the Sasquatch could be a close descendent of the Gigantopithecus. These, however, are merely theories and until the animal itself is proven to exist, all theories remain just theories.

In today's "enlightened" society man finds it quite easy to believe in the ghosts he can never see, the unscientific and possibly ludicrous ramblings of strange men with foreign beliefs but has difficulty believing what his fellow man swears to have seen, touched, heard or known. This type of objectivity regarding the Sasquatch, however, is not only necessary, but of extreme importance. In order to prove the animal is alive and is indeed an animal, no room can be left for frivolous acceptance. Such would hurt the progress of research and open the door to classifying Sasquatch with mythical creatures before proof can be found.

There can be no exact number showing how often this animal has been seen as it is quite possible that most sightings go unreported. Because the Sasquatch has been regarded as a "silly myth" for so long now, it is strongly believed that many witnesses will clam-up to fend off negative publicity and ridicule. However, at the time of writing, 349 reported incidents have been brought to my attention.

In the United States there has been 57 reported sightings in Oregon, 98 reported sightings in Washington and 66 reported sightings in Northern California. In Canada there has been 105 reported sightings in British Columbia (B.C.) and 23 reported sightings in Alberta. According to the numbers in my files, the Sasquatch, if it indeed exists, seems to be dwelling in the forests, mountains and costal regions of Western North America.

The 23 reported sightings in Alberta is what this book will concentrate on. Who are these 23 people? What was it they saw? Why did they report it? Where exactly did the incident take place? When did it happen? And finally, how do we know that they actually saw what they claimed to have seen?

In Alberta, a small part of the public school curriculum deals with "modern unproven monsters". The Ogopogo, The Loch Ness Monster, The Abominable Snowman and the Sasquatch. The Sasquatch seems to be presented as a single monster with a "boogy-man" type personality and the sighting of such equalized with that of a vampire, werewolf or a witch on a broomstick. The possibility of its existence as a rare, yet unfounded animal possibly harmless to the human race has all but been excused. This attitude has helped to place the animal along side fairy tale figures most children leave behind in their adulthood. Unfortunately, it has also led most adults to believe that there is no evidence of the animal and no serious research into the subject - What man in his right mind would spend time researching Peter Pan, Cinderella, The Wicked Witch of the West or, and especially, the "boogy-man"!

Common questions asked by people who discover that serious research is taking place are: "If it is in the mountains, why haven't I seen one? Why hasn't one been hit with a car? There are bear sightings reported and dealt with all the time, why hasn't the Sasquatch been given the same treatment?"". One of the most common questions is,"Why hasn't the bones or remains of a dead Sasquatch been found?" This is the toughest question to answer.

However, finding a Sasquatch in the mountains or forests would be like looking for a needle in a haystack. If it is there, the chances of finding it are quite rare. Professor Grover Krantz estimated some years ago that there may be only about 200 Sasquatches living in the Pacific Northwest.

The province of Alberta is larger than many countries. Much of it, however, is unpopulated mountain and forest terrain. Tourists, campers and hunters do run into bears and other wildlife but not everyday. Using the bear as an example, we will examine the possibility of running into a Sasquatch in the wilderness.

Manhunts, plane searches, war games, nature excursions and full tours take place in the foothills and Rocky Mountains of Alberta regularly. During these times, bears are rarely sighted. The bear population of Alberta is quite high, yet a person who lives and travels through Alberta for ten to twenty years may never see one unless they go to a zoo displaying the bears. If the Sasquatch does exist, it is obviously a rare animal who may avoid or fear human contact. If this be the case then 23 reported incidents involving the Sasquatch is not an unbelievably high number.

Of the sightings reported to me, each person involved was questioned extensively. If they had enough information to be convincing an interview would be granted to get to the facts and expose the fallacies. Many of these people have come forward in fear of exposure. They are convinced that they have seen a Sasquatch but are tired or afraid of ridicule. This is their story.

Author searching in Kananaskis country.

CHAPTER ONE
"Bigfoot And The Big Horn Dam"

On (July 4, 1986) I placed an advertisement in the classified section of the Calgary Herald requesting information on the Sasquatch. For years having researched, read, studied the subject of this animal and having a keen interest in the proving of its existence the ad was placed so that recent sightings in Alberta could be reported to someone and the story weighed for legitimacy.

Early afternoon of January 25, 1988 the phone rang. Debra Malone of Calgary, Alberta, was ready to let out a secret she held inside for sixteen years - she had seen, up close, an animal believed by her to fit the description of the Sasquatch. We set up a meeting.

Although I had many prank calls as a result of the ad, Debra quickly made it clear that she was determinedly serious and would not put up with being doubted. She was an outspoken friendly young woman who could not be classified as flakey in any way. She had, in spite of her own feelings, keep the Sasquatch sighting of her early teens relatively quiet. Those she had told interpreted the story as frivolous gesturing and conversational rambling. Her husband, who was possibly the closest to believing her, viewed the tale with skepticism.

Emotionally and often close to tears, Debra related what had happened in the summer of 1972 north of Abraham Lake on Highway 11 near the then in construction site of the Big Horn Dam in Alberta.

Debra, her younger brother Allen and her parents set out on the weekend to a secluded area near Nordegg and by Abraham Lake, just off of forestry road. During the drive they had passed the construction site of the Big Horn Dam. Past that point, however, the land was unpopulated, unexplored and undamaged by civilian occupation.

They continued driving on the highway for nearly 40 miles before turning into the bush. Debra recalls going about 20 miles further before stopping to set up camp. The area was dense with trees and berry bushes.

The day was still young and the family had made no specific plans. Debra was bored already. Instinctively attracted to nature though, she started towards the tree line and decided that now would be an opportune time for an exploratory hike in the woods.

She wandered through the natural mazes of tall pine and wild flowers taking in the freshness of the air and the untouched beauty. Eventually her wandering brought her to a bush of ripe berries and Debra bent down to pick them from the vines.

It was then that the stench hit her. An unfamiliar odor filled her nostrils and it was unlike any smell she had encountered before. Debra describes it as "...not as bitter as a skunk smell but muskier than that of a dog in heat". At that very same moment, the young city girl felt the presence of someone, something watching her. She whirled around, half standing, half squatting and found herself facing a creature who resembled no other creature she had ever seen or heard about.

Not more than 50 feet away in clear view and standing next to a thick line of bushes was an estimated eight foot tall hairy animal poised quite naturally in a perfectly human upright pose. *His arms hung down past his knees, his shoulders were at least three feet across and he was covered with long, matted, dirty brown and grey hair. The creature's nose jutted out from its face like a flattened ball of putty but it was his eyes which immediately claimed her attention - they were slate black, scrutinizing, even scary.

"That's no moose!" was Debra's first thought. She knew it couldn't be a bear... so just what was it?! Turning around and standing up her eyes met his. According to Debra it appeared so completely at home, so much in its natural

---

* *For the sake of grammatical clarity the male gender will be used in reference to the Sasquatch were no gender is evident.*

element that the possibility of it being otherwise didn't even cross her mind. He was standing up just like a person would but it was an animal. There were no solid trees in the immediate area, nothing to lean on or hold himself up by. He was not engaging in any activity, in fact, he just stood there staring right back at Debra.

For nearly ten minutes she watched him and he watched her right back. The creature made no motion of attack or sign of fear or even curiosity. He did not move a muscle and appeared to be undaunted by her presence. His eyes, she claims, communicated wisdom and instinct.

After what seemed like forever to Debra, the animal let out a deep throaty moan. Nearing a state of shock, she turned and ran, fighting her way through the trees, trying to remember the way she came; a wave of nausea hit, Debra started to vomit.

Finally finding her way back to camp, she burst through the tree line smelling of vomit, tears rolling down her cheeks. Her face went pale, she shook all over and was barely able to walk. Finding her brother she tried to explain what had happened.

Debra's mother heard the story and was convinced that her daughter had been spooked by a wild animal, possibly a bear. She listened, but did not believe that the girl had seen what was described - a description which fit that of a legendary mythical creature known as the Sasquatch.

Debra slept with her mom that night, still shaken and possibly in shock. Later, she was scolded for holding fast the story of "the big hairy man in the woods".

Not having previously known what a Sasquatch was, or having any former knowledge of the possible existence of such an animal, the sighting opened the door for Debra's education on "unproven animals". Approximately a year later she learned through a school assignment that such an animal is claimed by some to be in existence in North America. For the purpose of letting her story out, without facing more ridicule and scolding, she did a major writing assignment and presentation on book evidence and gathered knowledge of the Sasquatch. After the presentation, her instructor commented that: "it was written as if she had seen one herself".

Growing up, Debra would mention the Sasquatch sighting from time to time. In a circle of friends when the conversation turned to the unknown at a party or with a group of people she was close to, she would make the short and simple statement "...I've seen a Sasquatch".

She never gave a detailed account though because she was afraid of being laughed at.

Before she was married, however, Debra decided to tell her fiance about the experience. She, like most, was very reluctant to believe that the creature she encountered in 1972 was the same creature of myth classified with UFO's and unicorns. But, knowing her as he did it could not be denied that she had seen something and that she believed it was a Sasquatch.

Although Debra's story was told less and less she could not put it out of her mind. Eventually the experience invaded her dreams at night causing nightmares in which she would recall the experience as if it had happened that very day.

When at last, in 1988, Debra came across my ad in the Calgary Herald requesting information on the Sasquatch, she did not want to phone. It had taken some time for her to lift that receiver and dial the number but in doing so, a new chapter of her life would begin - she would not be alone in her experience. Others had too seen what she had seen.

In the course of my interview with her she admitted to wanting to see it again - only this time with someone else who could be a witness that the creature really did exist.

Is it possible that Debra Malone made up this story? That she carefully devised the tale in order to meet with conversational needs or have some claim to fame? Yes, such a thing is possible and has actually happened. The facts which substantiate her truthfulness, however, cannot be ignored.

Picture Drawn by Debra Malone, Age 13

- Prior to 1972, Debra had no knowledge of the existence of an unproven animal allegedly living in the unpopulated forest areas of the Pacific Northwest.

- Debra Malone allegedly was not fully aware of all the physical characteristics, of a Sasquatch reported, which so closely paralleled her description.

- Upon viewing the alleged photographs of the Sasquatch with me, she confirmed that it was the same breed of animal, then proceeded to describe, in detail the differences between the one she had seen and the one captured on film. The differences did not contradict any gathered (circumstantial) evidence, or the characteristics believed to be common among the alleged Sasquatch.

- At no time did Debra contradict, even accidentally, her original story of the sighting.

- Her sighting took place in an area where subsequent sightings have been reported by multiple witnesses.

- Although she is not a "shy" or "withdrawn" person, Debra has maintained a great deal of silence about the sighting for fear of ridicule, but has told those who are very close to her and in doing so has placed certain relationships in a jeopardy position in order to maintain that she did see what she claims to have seen.

- Excluding her description of its nose, the Sasquatch, as it has been reported time and time again, closely parallels the description she has provided. In order for her to gain information on these features, she would have had to do some extensive research.

- The stench of the animal is not a usually reported factor, when sightings are published, but the smell is an uncommon factor in most sightings and Debra's description of such matches both previous and recent descriptions.

A considerable amount of time has passed since my interview with Debra. Presently, she is collecting information on the Sasquatch, reading everything she can find about the subject and taking in every documentary. She is no longer afraid to "relate" the experience and now holds the attitude that if someone does not believe her, that is their choice. Her story, related and re-related has not changed a bit since the original interview. Her willingness to share the story, she claims, is a result of knowing that others right here in Alberta and more specifically in the area of the Big Horn Dam have also reported sighting a Sasquatch.

Her hope is that the animal can one day be brought, without harm, into captivity and documented as a legitimate species. She hopes that her husband will have the opportunity to see one and know without apprehension that her story is not a just a magnificent spun tale.

The Big Horn Dam, close to where Debra Malone claims to have sighted a Sasquatch, is the sight of the most publicized Sasquatch incident ever reported in Alberta.

Just past Banff National Park and West of Rocky Mountain House, by Highway 11 is the Big Horn Dam. In the summer of 1969 the dam was yet a construction site. Much of the surrounding area was retouched by modern man. Aside from the few men working in the area there appeared no evidence of human occupation in the immediate area - the land was considered "virgin land".

On August 23, 1969 a five man construction team working on the water pumping station encountered a Sasquatch. Harley and Stan Peterson of Condor, Floyd Engel of Eckville, Guy L'Heureux of Rocky Mountain House and Dale Boddy of Ponoka Alberta, noticed a distant figure on the high riverbank. They gathered together to watch it. According to Harley it looked like an incredibly large man. He described it as enormous, head slightly bent forward and very hefty.

Dale Boddy reported that it was too tall and it's legs too thin for a bear. As well, the speed it was moving at - approximate strides of 6 feet in length

- prompted him to believe that the creature was too large for any animal he had heard or known of.

Neither he nor the others with him had a pair of binoculars or a camera. They all turned to watch the figure. For nearly a half an hour the distant stranger stood stationary. Finally it sat down. Ten minutes later the thing stood up again hesitated for awhile then walked along the ridge of the riverbank and into the tree line. That is when the men lost sight of it. The animal had been in view for nearly 45 minutes.

Astonished by its apparent height and curious to know just how tall it was, two of the men headed to the riverbank to do a size check. When they arrived at the exact location, the three who stayed behind plainly realized that the creature who just walked off was more than twice the height of the men who just arrived, which would make it nearly 15 feet tall.

When the press got ahold of the story, witnesses to the Sasquatch animal came pouring out of the wood work. Mark Yellowbird, a Cree Indian also working on the dam had previously entertained his fellow workers with stories of human footprints along Rabbit Creek nearly 14 1/2 inches long and spanning more than six feet apart.

According to a Calgary Herald report August 30, 1979 (Bob Hage), Mark Yellowbird's daughter Edith along with three middle aged women had sighted "...four of the furry beasts working at something part way up a mountain along the David Thompson Highway".

Alec Shortneck, a worker at the dam, was clearing bush for construction when he reported to have sighted a strange creature watching him. He saw the figure about 50 yards away and was too stunned to do or say anything about it. Alec went on chopping at the bush and the figure left.

Vern Saddleback, a native in the area, claims that many people have actually seen the tracks yards away from his camp but never reported them or took the matter seriously. One Indian group, however, did report sightings. A band of Indians led by Chief Joe Smallboys were travelling in the area between Banff National Park and Nordegg in the spring and summer of 1969.

It was during this time many members of his band allegedly saw up-close, the Sasquatch.

In the opinion of the residents of surrounding towns and people on the local Indian reservations, a band of animals fitting the description of the elusive Sasquatch are roaming the area outside of Nordegg and near the Big Horn Dam sight.

The ridge at the Big Horn Dam in which workmen reported seeing a Sasquatch in 1969.

The late Chief Walking Eagle shared the opinion. He was willing to talk about it to his friends but never outsiders. According to The Edmonton Journal, August 30, 1969 (Nick Lees) the Chief was sure that his opinion would not be regarded as serious and he would thus be laughed at. So, he kept his ideas to himself and if he had any evidence of the existence of the Sasquatch, the knowledge was buried with him.

When the story reached full coverage Mr George Harris, a retired bulk fuel businessman, began making plans for an expedition to the Big Horn Dam site. Mr. Harris had, at one time, taken photographs of enormous footprints in the sand near the dam site which he believed were Sasquatch prints. In his report he stated that the prints could have belonged to a male and female of the species because of their varying sizes (13 and 17 inches long).

Gunter Schung, a guide and trapper from Nordegg, also made plans. He agreed to accompany Mr. Harris on the expedition. The two men were determined to track the creature down.

As would be expected, though, the expedition never took place.

During the press heyday with the Big Horn Dam sighting, a new angle was cast on the Sasquatch question: is the animal a revival of an Indian folklore legend used to scare the white man off from areas considered "sacred" by local bands of natives?

At the time of the sighting the Wesley Band of Indians were boycotting employment opportunities at the dam in a protest against the structure. The band's Chief John Snow spoke that his people felt that the dam would erase a sacred burial site and endanger his reserve located six miles downstream. Abraham Lake, which was formed by the dam, would undoubtedly ruin some of the land of the Kootenay Plains, destroy trap lines and game in the area.

It was because of these ideas that a journalist for the Edmonton Journal reported the possibility of Indian magic used to create some illusion in order to scare workers off the sight and possibly cancel out a ratifyed bill between the Provincial Government and Calgary Power Ltd. which solidifyed the construction of the dam.

It is possible, however, that all five men devised the story amongst themselves and set out to draw the attention of the media. Yet another possibility to be considered is "angles and the sun playing tricks with and obstructing a proper and clear view of an object or thing which could have been in motion at the sighting time".

The first possibility cannot be absolutely proven or disproven at this point. If the men were called together again and interviewed, first separately and then as a group, the time factor involved (20 years) would invalidate any changes or contradictions as it may be difficult to clearly remember all details 20 years after the fact.

Still, it is also possible that the story had not been true even if all five men remained in agreement to this date. If the men had lied to the media, the chances of coming clean now are quite slim. Also, it is relatively acceptable to believe that made-up stories are just as easy to remember as are stories of fact. Consequently, if this was not a true sighting it is twenty years too late to prove it.

Would it have been possible, though, for the sun to have hit such an angle of the earth at a tree or moving animal and cast a dark shadow or cause something to appear to be fifteen feet high? Yes, there are reported cases of the sun playing such a deception on man. In this case, however, it would be impossible. First there were five witnesses. Each of the men would have been standing at a slightly different angle than the other, thus one would be unable to see from the exact vantage point that another man was viewing the creature from. Secondly, all five men attested that the creature stayed long enough for them to observe it standing up, sitting down, standing up again and walking into the tree line. Evidently, the sun cannot accomplish a "trick angle" for the amount of time the men claimed to have watched the creature, nor could the sun cause such a trick to move across the ground and into the tree line.

Obviously, in 1969 at the Big Horn Dam construction site five men saw a creature standing, sitting and moving. The creature was over six feet tall and possibly as high as 15 feet tall. Was it a Sasquatch or a manifestation of Indian magic?

Authors research vehicle. A common sight on Alberta's back roads between 1984 and 2002.

CHAPTER TWO
"Needle In A Haystack"

The most common question asked of researchers is why, if the Sasquatch exists, hasn't the live or dead body or parts thereof been discovered? There are thousands of known animals in the world today, both rare and common, observed daily around the globe. The very idea of an animal that cannot be found and has never been proven by either the private citizen or the scientific community sounds completely absurd! Couple that with the knowledge that such an animal and similar creatures have been claimed sighted by one and more witnesses right around the world for many years but never brought into captivity adds even less credibility to the idea of its existence.

The Sasquatch, however, is most certainly not the only creature sighted but never confirmed as actual. Cryptozoology is the formal name for the study and research of unproven species. Such animals as the Ogopogo and the Loch Ness Monster as well as the Sasquatch have eluded capture despite ongoing research and upgraded testing methods. It would be easy to dismiss all the above as creatures of the imagination unless it is understood that the science of Cryptozoology is not a well staffed field; there are very few serious researchers looking into claims of creatures not featured in anyone's zoo.

While attending this year's conference on Cryptozoology, at the University of (Pullman) Washington I had the opportunity to meet a man extremely well known for his research into the Sasquatch. Rene Dahinden has been looking into reports and following up sightings for nearly thirty years. He co-authored the book "Sasquatch" (1973, Dale Hunter, Rene Dahinden) and has been credited with being the first person to devote himself completely to the finding of the Sasquatch. To this date Mr. Dahinden has not had the experience of actually seeing one.

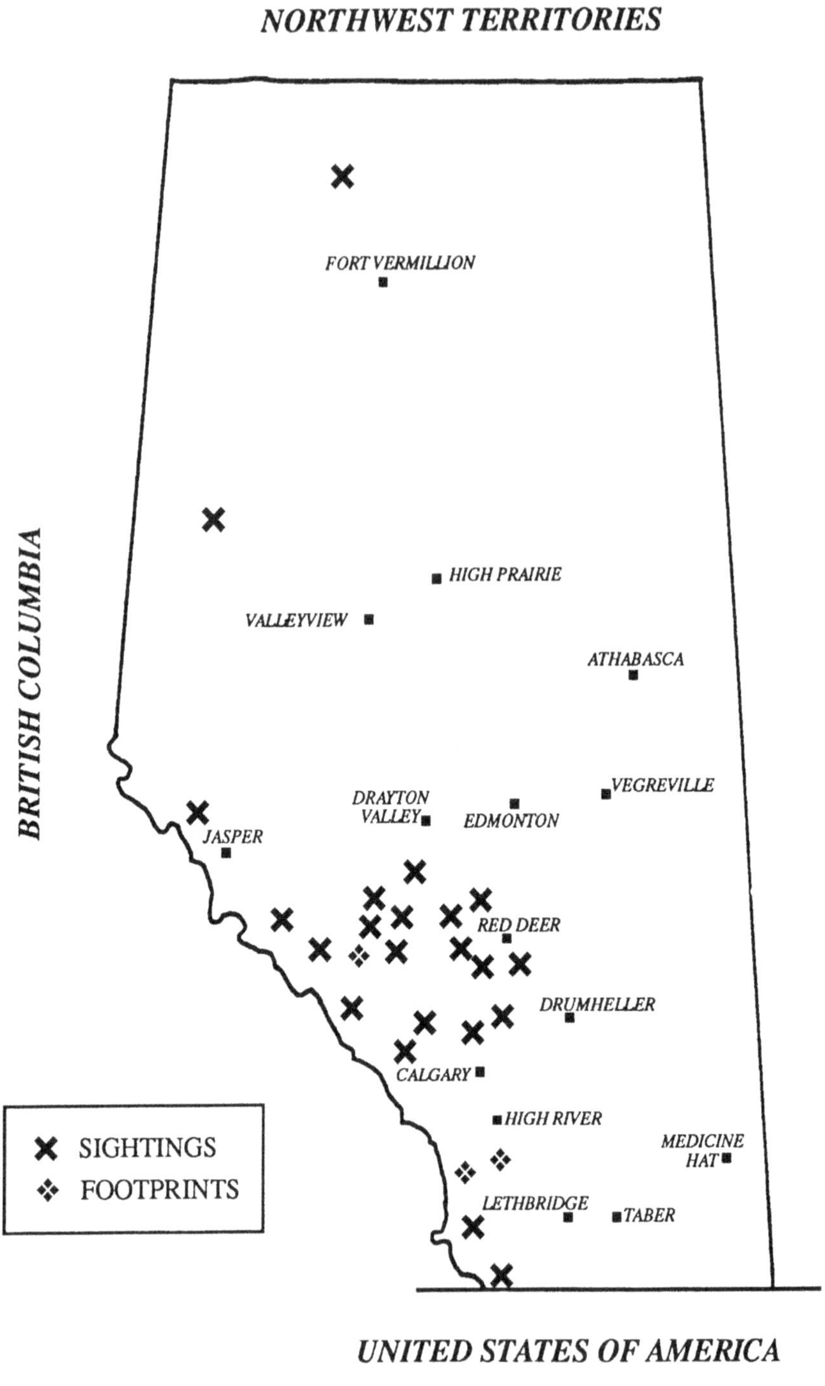

sasquatch in alberta

NORTHWEST TERRITORIES

✗

FORT VERMILLION
■

✗

BRITISH COLUMBIA

SASKATCHEWAN

✗

■ HIGH PRAIRIE

VALLEYVIEW ■

ATHABASCA
■

VEGREVILLE
■

DRAYTON
VALLEY ■
✗
JASPER
■

EDMONTON
■

✗ ✗
✗ ✗ ✗
✗ ✗ ✗ RED DEER
✗ ❖ ✗ ✗
✗ ✗
✗ ✗
✗ ✗ DRUMHELLER
■
✗
CALGARY ■

┌─────────────────────┐
│ ✗ SIGHTINGS   │
│ ❖ FOOTPRINTS   │
└─────────────────────┘

■ HIGH RIVER

MEDICINE
❖ HAT ■
❖ ❖
LETHBRIDGE
✗ ■ ■ TABER

✗

UNITED STATES OF AMERICA

14

With very few people working on the discovery of a Sasquatch or the confirmation of its non-existence, the fact that even common animals often elude sighting for long periods of time can add to the possibility that the right person has just not been in the right place at the proper time. To give an example of what I mean by common animals eluding sighting I will briefly take you back to the spring of 1986.

In 1986 I was serving with the army First Battalion P.P.C.L.I. and stationed in Calgary. On the day of our practise for the Regimental Ceremony The Trooping of the Colors, a small private plane crashed near a mountain in the Kananaskis region.

The Kananaskis area is a very widely used recreational park housing open fields, small towns, the foothills of the Rocky Mountains and miles of hills, dense trees and open fields. Tourists from around the globe visit Kananaskis on a regular basis and locals use it as a get-away from the whirl of city life.

When the plane was reported to have crashed, our government ordered the Air Force, the Army and the R.C.M.P. to begin a search of the area it was to have gone down in. We set up camp in Sibbald Flats campground and proceeded with a ground search. During the search 150 troops covered an area of 900 square miles of mountains, valleys and bush. The R.C.M.P., and the Air Force covered the area by helicopter and plane and several civilian volunteers also combed the ground. Hikers, campers, mountain climbers and a crew of people preparing Mount Allan for the 1988 Winter Olympics were also within the Kananaskis area. The plane, its pilot and passenger were not found until just before the search was to be abandoned.

During the search, the group I was with saw only one bear and it was from a considerable distance. Yet, we had covered an incredibly large territory known widely for its bear population!

Taking into consideration the vast number of people and equipment nearly unsuccessful in finding an airplane, its pilot and passenger in a Provincial Park - would it therefore be unreasonable to suggest that with only a few people looking and an area that is, essentially, the entire global surface of the earth, the Sasquatch search is not entirely unsuccessful? The Sasquatch search has yielded some results. Witnesses world-wide have sighted the creature, two men have filmed the creature and footprints have been found, photographed, reported and cast into molds.

Professor Grover Krantz of Washington State University estimated, as said earlier, some years ago, that there may be as few as 200 Sasquatch living in the Pacific Northwest of the United States. Professor Krantz is one of the few scientists who are convinced that there is something to the story of the Sasquatch.

Other members of the scientific community who are skeptical of its existence have stated that they cannot research something that has no claim to fact. Unfortunately, these scientists do have a point. If science devoted itself to proving the claims of a few unconfirmed reports of creatures not previously known, Vampires would be a major field of study.

On the other hand, the Sasquatch has been reported as far back as the discovery of North America. It was then that the Indians told the rest of the world about their creatures.

David Thompson, who was believed to be the first white man to find the tracks of a Sasquatch, kept a narrative (diary) which documented his encounter in 1811. One entry, dated January 7th was written about tracks he found in the snow while crossing the Rocky Mountains through the Yellow-head Pass near the present sight of Jasper. At the time, he was travelling with other early explorers as well as some Indians. A part of the narrative is as follows:

"Continuing our journey in the afternoon we came on the track of a large animal in the snow, about six inches deep on the ice. I measured it: four large toes each of four inches in length. To each a short claw; the ball of the foot stuck three inches lower than the toes. The hinder part of the foot did not mark well, the length of fourteen inches by eight inches in breadth, walking north to south and having passed about six hours. We were in no humor to

Author searching forestry road, Banff National Park.

follow him. The men and Indians would have it to be a young mammoth and I held it to be the track of a large grizzled bear, yet the shortness of the nails, the ball of the foot, and its great size was not that of a bear, otherwise that of a very old bear, his claws worn away. This the Indians would not allow." (Thompson, David p. 36)

Strangely, the Indians with him did not report the track to be that of a Sasquatch. This has led some to believe that it indeed was not a track of the creature. The size of the print, however, could not be easily identified with any other animal. The prints were reported to be 14 x 8 inches in size - not the shoe size of your average grizzly bear!

It would be safe to assume that a group of explorers passing through the Rocky Mountains would have encountered hundreds of animal tracks as well as many animals on their journey. Accepting the fact that there were no automobiles in 1811 and travel was done on foot or horseback, it would also be safe to assume that David Thompson and his men were travelling a vast animal-laden area for an incredibly long time (compared to modern travel standards). For Mr. Thompson to note the finding of tracks as an unusual occurrence he would either have to have lost his marbles, been exceedingly bored, or discovered a set of tracks with such rare qualities that they demanded both his attention and the time it would take him to note them in his narrative.

Footprint found around Baptiste River (northwest of Rocky Mountain House, Alberta) (October 1989).

David Thompson not only wrote of the tracks once, but on his return trip the following autumn he mentioned them again.

"I now recur to what I have already noticed in the early part of last winter, when proceeding up the Athabaska River to cross the mountains, in company with ... the men and four hunters, on one of the channels of the river we came to the track of a large animal which measured breadth by tape line. As the snow was about six inches in depth, the track was well defined and we could see it for a full one hundred yards from us. This animal was proceeding from north to south. We did not attempt to follow it, we had not the time for it and the hunters, eager as they are to follow and shoot every animal, made no attempt to follow this beast, for what could the balls of our fowling guns do against such an animal. Report from old times had made the head branches of this river, and the mountains in the vicinity the abode of one, or more, very large animals, to which I never appeared to give credence; for these reports appeared to arise from the fondness for the marvellous so common to mankind; but the sight of the track of that large beast staggered me, and I often thought of it, yet never could bring myself to believe such an animal existed, but thought it might be the track of some monster bear."

Although David Thompson's narrative is difficult to read and his grammar less than adequate by today's standards, the state of mind and his belief in what it was is impossible to ignore. Thompson believed that he had

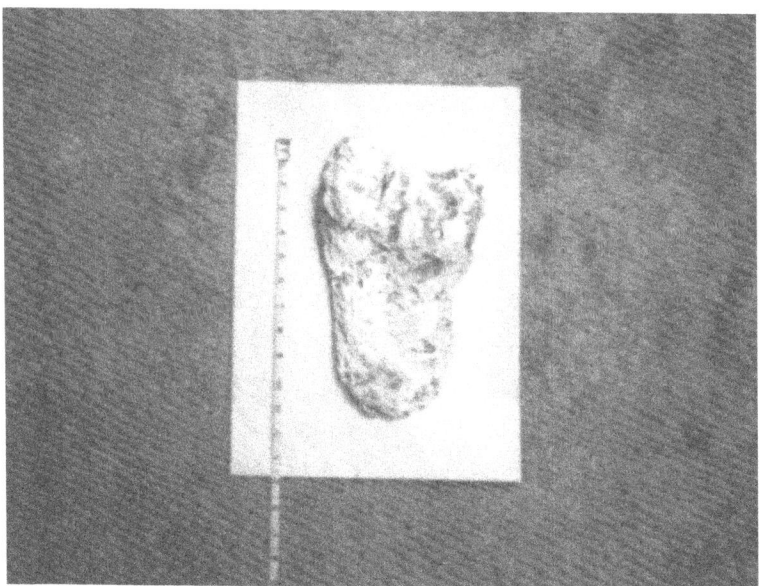

Casting of footprint found along Baptiste River, October 1989.

19

encountered the tracks of the beast which was commonly told of by the Indian peoples. According to Thompson, the tracks were large enough and deep enough to scare even his blood thirsty hunting party off from pursuit.

Thompson measured the size of the tracks with a tape measure, he did not guess at the size; then he recorded the shape, depth and on going length of the prints in the snow. Observing that these prints could be seen from 100 yards and were embedded in what he refers to as "ice" the animal who made those tracks must have been extremely heavy and incredibly large.

If Thompson and his men found the tracks of a bear the bear broke all records for size and set a new time record for a bear walking on its hind legs. Neither the hunters nor the Indians travelling in his party could adequately explain to Thompson's satisfaction what the species of animal was. Thompson had heard the stories, however, and by autumn had concluded that what he saw was the very same animal we now call the Sasquatch.

With David Thompson's report, the witnesses of the Big Horn Dam construction sight, footprints, hair samples and a film the only mystery left uncovered is why science has not yet begun a serious study into the question of the Sasquatch to determine, once and for all if such an animal exists.

CHAPTER THREE
"More Than A Legend"

In August of 1988, the Warden's office in Waterton Lakes National Park received a report from Darwin J. Gilles claiming that he, his girlfriend, along with a friend and his wife saw an unidentified animal the night before. The report read as follows:

> "At approximately 12:50 a.m. at the Crandell Lake Campground we spotted a very unusual animal. We were sitting at our campfire when we heard some snorting. We assumed it was a deer, but upon further observation we decided it was a bear and bolted for the cars. The animal was on its hind legs and we switched on the headlights on one of the vehicles. From the shadows I could see the animal was moving on its hind legs so I called to the other vehicle to turn on their lights.
>
> What we saw then was incredible. This animal was not only on its hind legs, it was striding (like a human) we watched as it walked through the trees for at least 3 to 4 seconds.
>
> I immediately thought it was a joke. We're all convinced it was not a bear. We jumped into the same vehicle and followed in (its) general direction...
>
> We came across another vehicle and flashed our lights. These people had also sighted something very strange and were quite scared. This confirmed that we had all seen something.
>
> It is important to note that we are four mature, responsible and professional people. We thought very carefully before coming in to report this incident at the warden's office. All four of us are convinced that it was not a bear. I am equally convinced it was

not a practical joke. If it was, it was pretty elaborate and well done.

From our sighting, the best description we can give is as follows: the animal was approximately 8 feet tall (as measured by the tree it was standing beside in our campsite). The animal was never on all fours. When we switched on the headlights and got a good look, this thing was striding... It also had long arms which were swinging while it moved through the bush. It wasn't a bear, okay.

I don't know what more I can write about this incident. We would appreciate hearing anything that might explain what we saw (or additional sightings, if any).

Darwin Gilles"

On the night of August 29th, 1988, Mrs. Susan Stoness, the first member of the Gilles party to sight the creature, responded to my newspaper ad for information on the Sasquatch.

Susan told of the incident, which involved her, her husband Scott Stoness, Darwin Gilles and Shannon Senkow, while on a camping trip in the Crandell Lake campground located in Waterton Lakes National Park in the southwest corner of Alberta. We proceeded with the interview.

On the evening of the sighting the two couples were attempting to play a game of hearts (cards) on a picnic table next to the fire pit by their campsite. The wind picked up, however, and blew the cards off the table scattering them across the ground. At approximately 12:50 a.m. they gave up the game and decided to turn in for the night. Susan and Scott headed down a path towards the public washrooms to brush their teeth.

Within moments Susan was frightened by a noise. Scott, knowing his wife was afraid of being in the wilderness at night, paid no attention to her complaint, took her by the hand and continued down the path. Then, at the same time they both saw a strange creature standing on the trail ten feet ahead. The animal made a low grunting noise. Susan yelled "It's a bear", tore away from Scott, ran back down the path toward their campsite. Darwin and

GORILLA FOOT

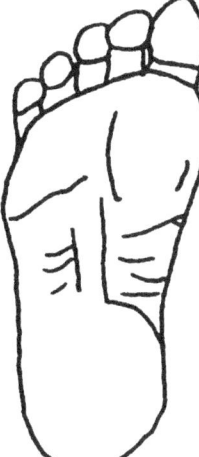

SASQUATCH FOOT

HUMAN FOOT

Shannon, who had not yet left the firepit heard Susan yell, stood up and turned around in time to see her flying back down the path toward them.

Panic struck. All three headed for the two cars. Shannon fled to one car while Darwin and Susan headed for the other vehicle. Meanwhile, Scott was still on the path, ever so slowly backing away from the creature. The creature moved off the trail and into the trees. Scott made it back to the campsite and ran for a car. He ended up in the same vehicle as Shannon - his wife would not open the car door to let him in.

Darwin sat in the car with Susan trying to keep her calm. Susan, shaking with fright, worried about other campers wanted to "lay on the car horn to wake and warn the neighbors". Darwin stopped her.

After awhile Darwin turned on the headlights. He saw nothing. Believing that the danger had passed he rolled down the window and yelled for Scott, whose car was facing the opposite direction. Just then an eight foot high thin hairy man-ape-like creature strode into the area lit by the headlights. In four strides it covered the distance illuminated by the lights, not running, just walking ... at an incredibly fast pace. Its entire body was covered with black hair. Its arms were much longer than the arms of a human. In a matter of seconds the creature had passed through the light and back into the trees, not to be seen again.

It was then that Darwin and Scott knew what it was they had all just witnessed - the legendary, apparently fiction animal known as the Sasquatch. The two men wanted to get out and search for tracks but Susan and Shannon were too shaken, they just wanted to pack up and get out. After some insisting, however, a compromise was made and the four campers drove around the area trying to spot the creature again.

In their effort to get another glimpse of the creature they came across a truck load of people who reported having seen "something strange" on the grounds an hour previously. The people in the truck did not give their names and no further reports were heard about the incident from these people.

After a night of restlessness and occasional sleep the Gilles party contacted the warden's office who followed them back to the area of the sighting to search for tracks. The warden kept asking if there was a possibility that the creature had really been a bear. Not one of the four witnesses were willing to concede to a bear story. In their words "It didn't look like a bear, it didn't walk like a bear, it didn't even resemble a bear". There was just no evidence to substantiate the creature being a bear.

Susan coerced Darwin and Shannon to take part in the interview at her and Scott's home. Not surprisingly, the four witnesses had more questions to ask of me then I did of them. Each person was interviewed alone in a separate room and asked for their version of the sighting.

Following are excerpts of the interviews conducted:

Susan Stoness:

Q: Describe what you saw.

A: *(I saw) a creature first from about 10 feet away in the outskirts of the campsite, located approximately 15 miles outside of Waterton. My first reaction was that it was a bear. He grunted as we approached. He was standing on two legs and covered with black or dark brown hair. I would estimate the height to be eight feet tall, (in accordance to the height of the tree we later measured against).*

He was heavy, huge, maybe six to eight hundred pounds. His face was flat with eyes, nose and a mouth. The arms were very long and the tips of his fingers came very close to the knees.

Q: How long, in total did you see the creature?

A: *Approximately one or two minutes.*

Q: When you saw the creature did it react in any way, and if so, exactly how?

A: *He grunted at us ... Not seeming hostile or scared. (I think) he was letting us know that he was there, almost like a warning.*

Q: In your own words describe what happened.

A: *...My husband held me by the hand and as we started down the trail I thought I heard something and I told him, "I think I hear something". He really didn't take notice of that because I always hear stuff; I'm kind of scared at night.*

We took a couple more steps and then we saw a big hairy creature standing up in front of us, probably ten feet away. Then it grunted at us three times. Well, I screamed, "It's a bear," and ran to the car. Darwin got in the car with me, while Shannon ran into the other car. Scott, when he finally got to us, tried to get in the car with us but I wouldn't open the door so he ran to the other vehicle and jumped in with Shannon.

... The two cars were bumper to bumper and when we put the headlights on and looked around Darwin thought he saw something moving by the fire towards the trail. We yelled for Scott to turn on his headlights. About 10 seconds later the creature walked into the light. It was walking on a ridge about 30 to 35 feet in front of the headlights. It looked back at us but did not break it's stride. It wasn't really fat like some of the pictures we have seen, in fact it was slender with really long legs, disproportionate to the body. My husband yelled "holy ... that's incredible, it's a Sasquatch".

We drove around the campground and came to another truck full of people flashing their lights at us. The four people in the cab of the truck listened to our story and claimed to have seen something similar about 20 minutes earlier, but they were quite drunk and only three of the four actually saw it. When we got back to our campsite I did not want to get out of the car or stay the night in that place. Scott got into the car with me and fell asleep. I stayed up all night and watched the trees.

Darwin and Shannon stayed up all night too by the fire. The next morning, after much debate, we decided to report it to the warden. We went to his office and waited for him, told our story and Darwin gave the report. The warden tried to convince us that we had seen a bear. He kept saying *"well, bears stand up on their hind legs, are you sure it wasn't a bear?"* He was nice, but I think he thought we had seen a bear.

Scott Stoness

Q.   When and where did this sighting take place?

A.   *The incident occurred in Waterton Park, Crandell Lake Campground. It was the May long weekend, early (before 1:00 a.m.) on Monday.*

Q.   Could you describe the creature?

A.   *It was walking on two legs, I never saw it go down on all fours. Its hair was dark brown or black and it blended into the darkness - I didn't actually see it until we got very close. The creature was between 7 1/2 and 8 feet tall and approximately 500 pounds.*

The second time I saw it, while in the car, I noticed its stride was about 5 feet. It moved very quickly through the beam of the headlights, keeping its legs almost straight.

Q.   Your wife mentioned that it made a noise. Could you describe the noise?

A.   *The sound it made reminded me of that of a bull when he is in pursuit. I was chased by a bull once and the sound it made is the closest I can come to describing the creature's noise. I do not think that any human could have made such a sound - it was like an animal with a big throat blowing a lot of air.*

Q.   Once Darwin filed the report, the warden returned with you to look for tracks. What was found?

A.   *Well, the area was very rocky and covered in moss. I think an elephant could have trampled through unnoticed. We did not find any tracks to substantiate the sighting, but considering the terrain I am not surprised.*

Q.   At what point did you believe this creature was a Sasquatch?

A.   *The second time I saw it I was pretty excited because I thought "this must be what everybody says is a Sasquatch. Both Darwin and I wanted to chase after it and see what it really was but my wife wasn't too keen on the idea. Instead we drove around awhile. A big truck with 3 or 4 people in the front approached flashing their lights at us. They asked if we had seen anything. So we described our creature and they described theirs. They also thought the creature was a Sasquatch. I don't think they saw it as well as we did though. When we told the warden he tried to convince us that our creature was a bear.*

Darwin Gilles:

Q.   When and where did this incident occur?

A.   *On May 23 at 12:50 a.m., site C-3, Crandell Lake Campground, Waterton National Park. It was a typical campground with a circular gravel road. We were on an elevated site about three feet up from the main path.*

Q   How far were you from the creature?

A.   *When I first saw it I was roughly 20 to 30 yards away.*

Q.   What do you think this animal was doing in the campsite?

A.   *Initially, when Scott and Susan left the campground walking towards the creature it grunted. They thought it was a deer and continued walking, then it grunted again. I believe that the creature was just curious and I' m convinced that it was watching us for quite awhile. The reason being is when the cards blew off the table about 15 minutes before Scott and Susan saw the animal, I remember hearing something. I am convinced that it was just curiously watching us.*

Q.   From what you remember, describe what happened that night.

A.   *We were playing cards by the campfire and Scott and Susan decided to turn it in. They started off down the path to brush their teeth. As soon as they walked out of the campsite we heard the first grunt and I jumped up. Scott yelled back "it's just a deer" and grabbed Susan by the hand continuing down the path. Then this thing grunted again. This time*

Susan said something like "that's not a deer, its a bear!". As soon as Shannon and I heard that we were up. Shannon suggested we move calmly towards the cars but it was like a free for all, we sprinted. In the confusion, I ended up in the same car as Susan; Scott ended up in the other car with Shannon. Our car was facing inward toward the bush and away from the direction of the first sighting. Susan was quite excited and wanted to honk the horn but I stopped her and turned to look back in the direction of the bear. Then I saw a tall skinny shadow. "What is a bear doing on its hind legs" I thought, "usually they come down off their hinds if they are standing up". I turned on the headlights but couldn't see anything, so I rolled down the window and yelled for Scott to turn on his. The four of us saw this thing plain as day. It was leaving our campsite walking on an angle away from us across the beam of the headlights and into the trees. There is no mistaking what we saw.

Shannon Senkow:

Q. Can you remember what your initial reaction to the sighting of this creature was?

A. *Yes. I thought it was a bear so I was scared and ran for the car.*

Q. When you finally did see this alleged bear, what did it look like?

A. *It was standing there covered in dark hair. It had very long and hairy arms and stood a good 7 or 8 feet tall. With the headlights of the cars on I saw that it wasn't a bear. It took about ten steps in a human-like gait and went on its way.*

Four people saw something in the middle of the night far away from the city they lived in. Is it possible they could have seen a bear? Taking into consideration these four people are professional, intelligent careerists I personally believe that it is ludicrous to assume that they do not know what a bear looks like!

The only possibility of these people having not encountered a Sasquatch is that they were victims of a clever and well prepared practical joke. The four people in a truck could have played such a prank and returned to the scene to see if their theatricals were effective. This in my opinion is the only other possibility and giving consideration to the size, stride and noise of the creature, the practical joke theory does not hold much weight.

The life size statue of a Sasquatch at the Museum of Natural History in Banff, Alberta. The general body proportions seem to be correct, however, I disagree with the skin colour on the face and hands.

At the time of writing, I am trying to contact the warden who received this report and went looking for tracks in order to verify that no further evidence was found.

The "Legend of the Sasquatch" which began as an Indian Legend and mountain region folklore has evolved into a serious mystery which is gaining in credibility by the day. In the eyes and ears of intellectuals it is one thing to hear of distant tales told by distant people but it is yet another matter to have these tales substantiated by other intellectuals who appear to have no personal gain in adding fact to what was once mere fantasy.

Campsite C-3, Crandell Campground, Waterton lakes National Park.

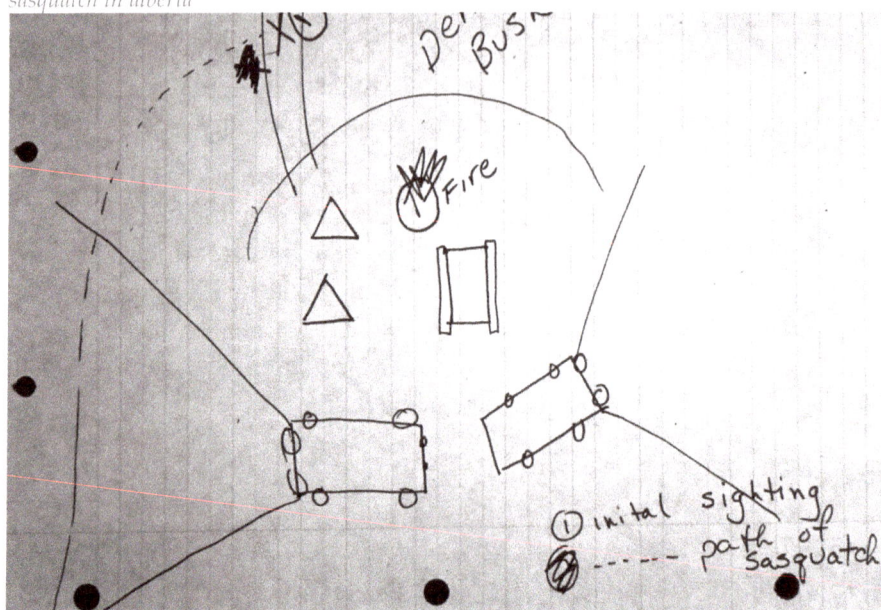

Upper: Witness, Susan Adams drawing for the author of the campsite and event.

Right: Map of Crandell Campground.

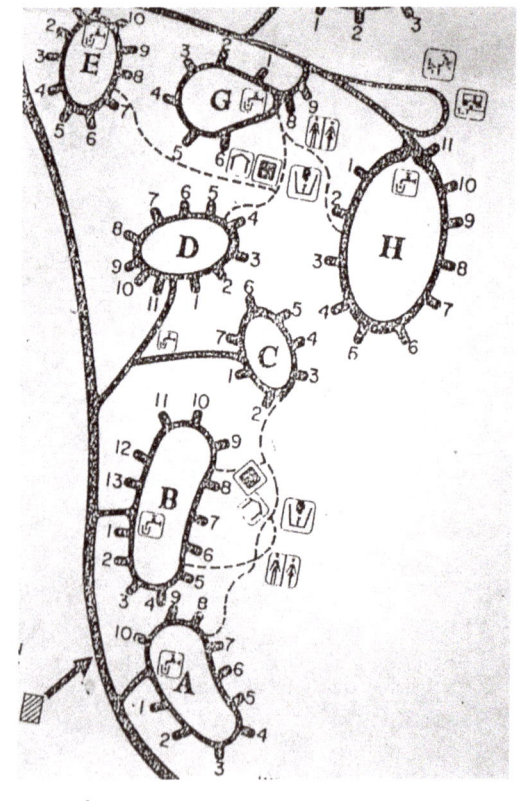

CHAPTER FOUR
"The Tallest Man On The Mountain?"

As reports of the Sasquatch become more and more frequent the question of height raises more and more concern. Reported sightings of the creature are almost contradictory in this area, giving the animal's height from a mere 7 feet tall to a whopping 15 feet tall. Both those who have made height surveys and those who guess verify that an extreme difference in height exists.

Investigators of the Sasquatch most often assume the animal to be between 8 and 9 feet in height. In the past, reports that placed its height in the area of 12 and 13 feet tall were regarded as excited over estimates made by those who were too surprised to give an accurate record.

From Northern California to Washington State and up into British Columbia, the most frequently reported height is between 7 and 9 feet. Many Alberta reports also fall into this height category. In Alberta, however, an increasing amount of sightings claim that the creature stands upward of 12 feet. Nearly one quarter of all Alberta sightings have placed the creature into an unusually high height category.

One sighting claims that the entire head of the animal could be seen above young 9 foot trees. The Big Horn Dam incident placed the height at 15 feet. The young man cutting wood with a band of Indians led by Chief Joe Smallboys saw a 12 foot tall creature at a distance of 300 yards. The two creatures sighted by a man on Highway 11 at a distance under 200 feet were reported to both stand over 12 feet.

Do these contradictory sizes take away from the possible authenticity of reports, or could it be that the Sasquatch is no more unique than the human man in this area? The average height for a man is 5'11" - yet we have a large population of men over 6'5"! The NBA boasts of players over 7' tall. Could

**A six foot two inch man,
in comparison with an 8 foot Sasquatch.**

it be possible that the Sasquatch reaches the height of 8 to 9 feet tall and the female of the species grows to a height of just over 7 feet - with the odd individual reaching heights of 12 feet and over? I personally believe that this is quite feasible. But why are so many of them seen in Alberta? Of this I could only guess - and guess I think I will... A possible reason for the Sasquatch reaching such unusual heights in Alberta could be a mere coincidence.

It is true that in British Columbia reports of 12 foot and over Sasquatch are quite rare. It is also true that the terrain and atmospheric conditions are quite similar across the border of Alberta and B.C. However, B.C. reports far out number Alberta reports and by coincidence Alberta sightings could have just been, thus far, of taller animals than those of our neighbors. It is also important to take into consideration that the excitement of the moment could cause inaccurate guess work on both sides of the border.

There is, however, very little known about the Sasquatch at this time. Until the animal is discovered and researched the question of height will remain a mystery. Possible reasons for the differences can only be guessed at in light of what man now understands of the adaptability of animals in general.

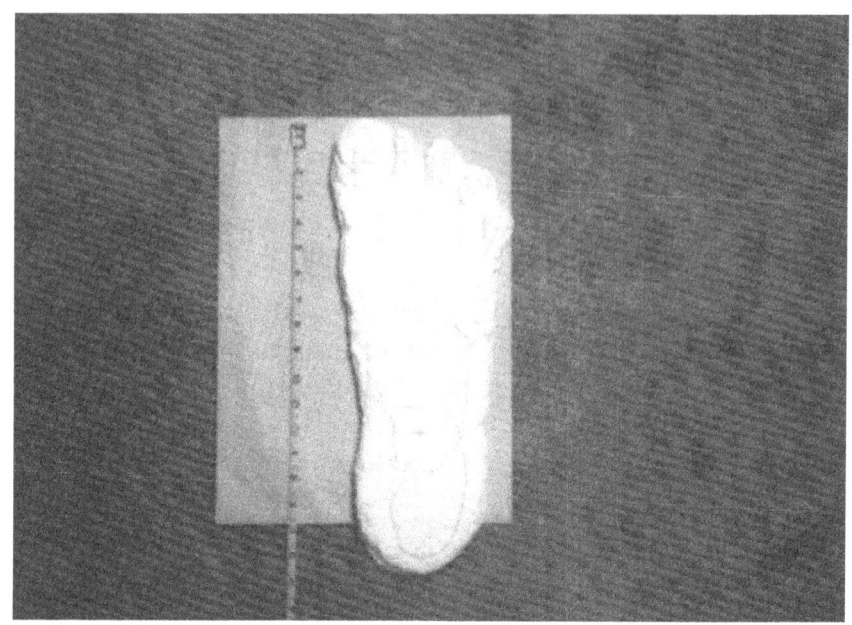

Cast of Bossburg Cripple (left foot). Found in northeast Washington State, 1969/70.

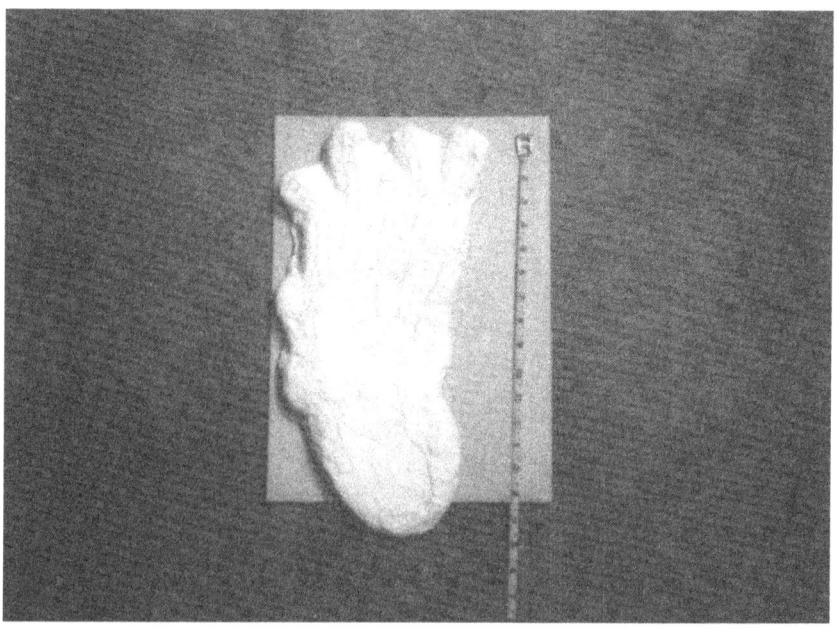

Cast of Bossburg Cripple (right foot). Found in northeast Washington State, 1969/70.

CHAPTER FIVE
"Should There Be A Doubt?.."

In the years that the Sasquatch animal has been pursued, it has become evident that its natural habitat is among the trees, in the mountainous regions and away from civilized man. Still, there has been reported sightings of the creature wandering onto private property near and just beyond the cities, ignoring thus far observed caution and coming dangerously close to civilization.

Reports of the Sasquatch peering into windows, approaching camp-sites, even opening the flaps of a tent left unzipped are on file. Incredibly, there are incidents known of where the animal has run beside a moving car (possibly to get a closer look at the driver). Some have been known to shake vans, trucks and campers. Owners of these vehicles have heard noises at night and found large human-like footprints encircling their vehicles the next morning. Is it possible that the Sasquatch is a Curious-George who stalks the night searching and exploring?

On the night of November 26, 1987, at approximately 8:30 p.m. a 14 year old girl phoned me to report a sighting which involved her and a friend, that took place outside of Airdrie Alberta the previous evening. My own Curious-George nature got the best of me as Airdrie is mainly flat prairie land sparse in trees and even lacking enough brush to hide a jack-rabbit for any length of time.

The girls, who did not wish to be identified in publication, claimed to have seen a Sasquatch wandering around. For the purpose of clarity I will identify the caller as Jill and her friend as Ann.

The two girls had climbed a high fence on the extreme western edge of town that overlooked the railroad tracks. On the other side of the tracks was a smaller fence about 3 feet high. Beyond the fences the girls had a clear view of farmers fields and the distant foothills and mountains. Both Jill and Ann were forbidden by their parents to cross over the fence, and in the spirit of youth, they crossed over it anyhow. Ann climbed over the larger fence, crossed the tracks, climbed an embankment and started for the smaller fence. When Ann had reached the second fence, Jill started her climb. Halfway up she heard Ann screaming and saw her flying back over the tracks yelling "Let's get outta here!"

Both girls scrambled back over the fence and ran into town. Jill confused by the panic finally stopped Ann and asked for an explanation. After Ann calmed down she shared a story with her friend that terrified and excited both of them.

As Ann was about to climb the smaller fence she heard a noise in the farmer's field ahead. Approximately 100 feet away a large black man-shaped creature stood up from a crouch to an incredible height His whole body was covered in hair and he was taller and wider than any man she had ever seen.

Jill, still climbing the first fence had seen nothing but knew by the degree of Ann's fear that her friend had seen something.

The next day Ann was still in a state of fear. Later that day when Jill phoned me to report what had happened I could hear Ann in the background. She really didn't want to talk to me and was not too pleased with the fact that Jill did.

Myself and two others paid a visit to the site the next day. We could see the foothills in the distance and estimated them to be approximately a half a days walk. The ground was still frozen though, and we found no tracks or evidence to back the girls story.

Common sense dictates that a story like this should be viewed as a possible prank. According to Jill no one else knew they were out there so it is not reasonable to assume that a third party played a prank on the girls. Instead, a sensible question would be "Are the girls playing a prank on Sasquatch investigators?"

Forest Regions of Alberta

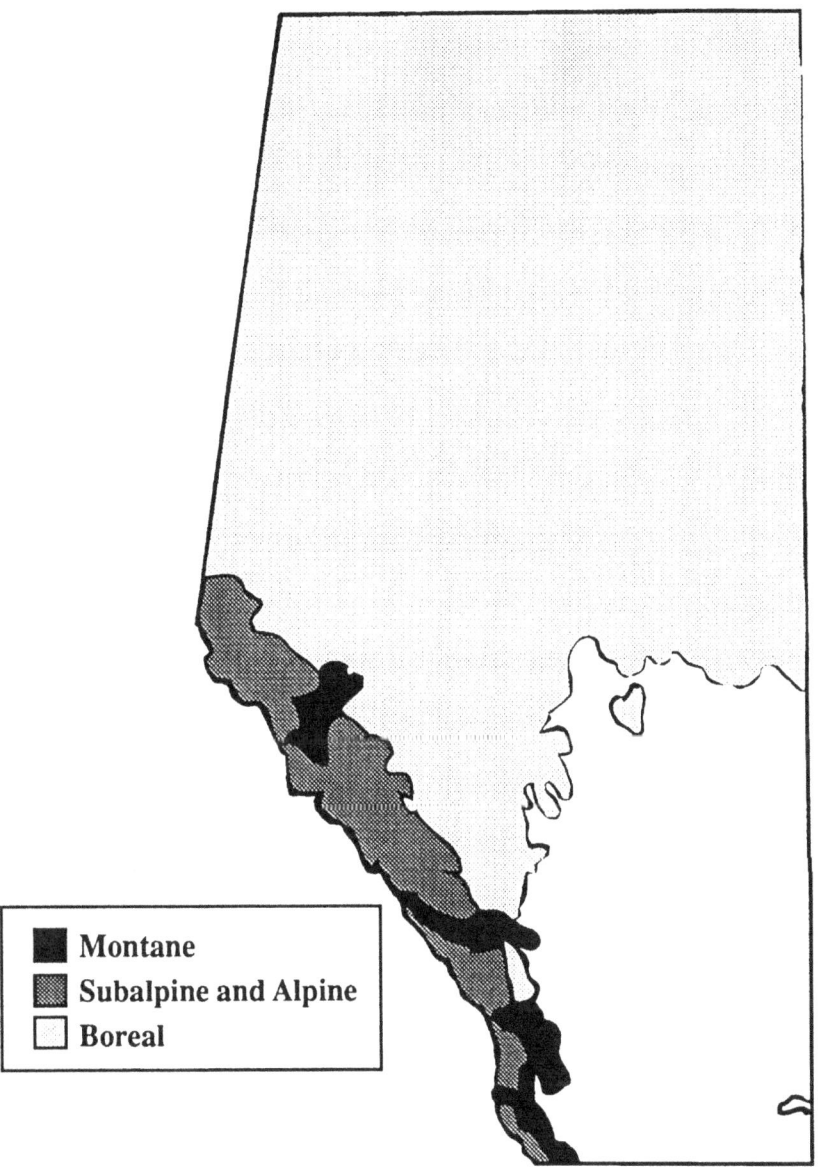

Montane
Subalpine and Alpine
Boreal

Taking into account their age, the fact that they did not want to be identified and of course their immediate response to a newspaper ad in the classified section of a neighboring city it would seem quite possible that two girls were simply looking for adventure and attention.

It is also possible that Ann had seen a human man and fear and imagination created the rest. The factor which lends credibility to the story is simply that Ann was the only witness and her description was not exaggerated. If she were seeking attention or adventure, in my opinion, the actual sighting would be more detailed.

Ann and Jill's story is not the "closest to home" account I have on file. In 1978 there was a reported sighting in the Bearspaw area just outside of Calgary. Bearspaw is located directly above the north bank of the Bow River. From the high bluffs, there is a spectacular view of the Rocky Mountains a mere 35 minutes away. The area is dotted with beautiful homes, small ranches, hills and trees along the river bank. The river is upstream from the town of Cochrane and passes through the Stoney Indian Reserve and into Banff National Park. If you were to walk along the Bow River in this direction you would find yourself in the foothills with the Alberta Prairies lost long behind the thickening tree line.

Because of the close proximity to heavily wooded areas Bearspaw is an ideal location for a Sasquatch to come close to the city undetected, and slip back into the wild in less than a day's journey.

Ann and Jim Smith (alias) called me from their home in Cochrane to report the Bearspaw incident. The mother and her son had lived in Bearspaw at the time of the incident.

They both agreed to an interview. Ann and Jim were kept separate during the questioning and while Ann was giving her answers my friend Robert Alley drew a sketch of the creature they had seen under Jim's direction. Unfortunately, the two stories were contradictory and full of holes. On the presumption, however, that time fades memory and that two view points can

Sketch of the creature seen by Jim Smith, drawn by Robert Alley under Jim's direction.

create confusion, I have included excerpts of the interviews to allow the reader to decide for himself.

Ann Smith

Q.   Where and when did this incident take place?

A.   *The incident took place either in '77 or 78 (I believe it was 78), approximately 3 1/2 miles northwest of Calgary towards Cochrane. It was between 9 and 10 in the evening on a scarcely used trail leading down towards the river.*

Q.   How far were you from the creature and what was your initial reaction?

A.   *I was approximately 1/8 of a mile from the animal. At first I tried to figure out what it was. I thought it might be man with a pack sack or carrying a large sack or one of the local boys going home. But it was too square to be a man.*

Q   What was the appearance of the creature?

A.   *It was standing upright on two legs, very sturdy legs and was covered with black fur. It was 7 feet tall, maybe 300 pounds and its arms did not swing when it walked. It didn't make any sound but it smelled horrible.*

Q.   Did you check for tracks or any evidence and did you report this to anyone at the time?

A.   *I was afraid to check for footprints and so was my son. We didn't report it to the police or anyone because we were afraid of being laughed at. When we told family members they got a kick out of it.*

Q.   Could you try to remember exactly what happened and describe the events?

A.   *The dog started to whimper outside so I figured that some other dogs were picking on him. I went outside and he was under the porch whimpering in a way that I never heard him whimper before. Then the odor struck me. It was awful. I called my son and we got the dog out and continued with what we were doing. When I went over to the french windows I noticed that it was snowing and then I saw this thing walking in the direction of the river. That would take him across the Bearspaw road like in that area down the Transalta Incemanation Station. It was walking very quickly. I called to my son and said "Could you see a packsack on his back?" because it looked so very big. My son said it looked like it was covered in black animal fur. Its legs were very stocky, it took big strides, it was very squared in the upper body area and it just kept walking until it was out of sight.*

Ann's story sounded quite authentic and so did Jim's, only their stories sounded like two different sightings! When I had completed the interview with Ann I called Jim into the room and this is the version he gave...

Jim Smith

Q.   When and where did this incident occur?

A.   *November, 1978 in Bearspaw around 10:00 p.m. The area was hilly with flat grass and stubble grass. I was about 200 yards from the creature.*

Q.   Could you please describe your initial reaction and what the creature looked like to you?

A.   *My first reaction was that it was BIG, real BIG! It was walking on two legs with the upper part of the torso hunched. The creature appeared to be covered in black hair, standing approximately 6 to 7 feet in height and about 280 pounds Its arms hung to its side and there was a smell like sulfur.*

Q.   Did you check for footprints or report the incident?

A.   *No and No.*

Q.   Would you describe exactly what took place.

A.   *At about 10:00 p.m. my mother and I were sitting in the house and the dog started to bark. I went outside to see that the problem was. He was a guard dog so he only barks when something is in the yard. I went out to the porch to calm him down and dog shot into the house whimpering. I started to smell this stench and I looked up across the road. I saw this huge figure and knew it was too big to be a person. I went back into the house and onto the front porch and we watched it walk in a southward direction toward the river.*

Q.   Did you see it first or your mother?

A.   *I did.*

Q.   Then you told her and she came out to watch it with you?

A.   *Yes.*

Both Ann and Jim claim to have seen the creature first. Ann through the french doors and Jim on the porch. Ann claims the dog was whimpering under the porch and she went outside to see what was the matter with it. Jim's story makes no mention of the dog being under the porch or Ann going out to see it. Jim described the upper torso as being hunched, Ann saw it standing very straight. Ann saw it with Jim for 4 to 5 minutes while Jim saw it with Ann for about 2 minutes.

Either one or both of the alleged witnesses has forgotten the details of the event or there was no event. They both did agree on the sketch and the details. Their given reason for withholding their true identities is that they claimed to have taken enough ridicule from the family already.

I cannot confirm that this was an actual sighting of the Sasquatch and I cannot prove that it wasn't. If, however, it was the fact that they both agreed that the sketch was the creature in question, a sketch of a Sasquatch, and I know that the Calgary Zoo did not lose one of its gorillas that day, it would leave out the possibility of any other known animal. So, it is up to the reader to decide if this incident warrants the credibility of an authentic sighting, a simple case of fiction or that of mistaken identity. And, despite the time between the event and the interview (9 years) the discrepancies cannot be ignored.

CHAPTER SIX
"Sasquatch Drinks Coors"

As an advertised investigator and researcher into the strange and often humorous subject of the Sasquatch I have received more than a shareful of prank calls, jokes and just plain rude and nasty reports. The jokes and prank calls often lend a side of humor on a subject with very serious applications and implications.

One night I picked up the phone and the caller on the other side identified himself as a having information on the Sasquatch. "When and where did you see the animal?" I asked.

"Oh, I saw him about right now, in fact he's sitting here having a brew with me, want to talk to him?"

Plenty of midnight callers, with nothing more to do than bother others have dialed my number and claimed to see the Sasquatch. Sometimes the stories involve partying with the animal or watching it descend from a UFO. If the animal is a native earthling, the idea of an intergalactic heritage is a little far out into left field (or left galaxy, if you will).

One man (who has, thank heavens, never dialed my number) claims to communicate with the creature using ESP. Other claims made place the Sasquatch out of the realm of reality and into the realm of fantasy. The only story I have yet to hear is that of a Sasquatch riding through the sky on a unicorn.

Those who make these claims and some who go as far as to state them publicly are known among researchers as "the lunatic fringe". Many of this classification have engaged in writing documentaries and claiming authority or sole knowledge of the animal. It is important to note that those in the lunatic fringe have little to no evidence to back up their claims to authenticity or authority. If the Sasquatch is indeed a creature or animal living in some areas of the world, no one man could have complete authority on the subject unless he alone had all the evidence and a Sasquatch to back it up. The subject of this animal is yet on the threshhold of science and will remain there until a body with bones, sinew, tissue and muscle is brought to the inner circle of science and revealed to the outer circle of civilization.

This is one example of how the Sasquatch has become a local tourist attraction in some places. The sign reads "Big Foot Campgrounds", and is located at Harrison Hot Springs, B.C.

Even the B.C. Government has gotten into the act. This Provincial Park is located just outside of Harrison Hot Springs, B.C.

First sketch drawn by witness
known only as 'An observer' of
creature seen outside Medicine
Hat, December 1973.

CHAPTER SEVEN
"What Is The Sasquatch?"

The text of this book has, thus far, covered many sightings and possible encounters with the Sasquatch. What is it exactly, though? With the combined effort of other researchers, years of eye witness testimony and the few theoretical possibilities from the scientific community, a composite-type verbal description can he formed dealing with physical characteristics and probable instinctive/habitual traits or practises.

Based on current evidence, it would be reasonable to infer that the Sasquatch is a higher primate, hominoid, which is a classification reserved to describe mammals in the man-like family such as apes, orangutans and humans. Presently there are only five identified higher primate species, the Sasquatch, if proven to exist could possibly be the sixth.

According to eye-witness testimony the animal moves only on two legs and, although it walks very much like a man its stride is nearly twice as long and its speed is considerably faster. Scientists examining a film made of the creature claim that it would be virtually impossible for a human person to imitate the stride and speed, even with extensive training.

The Sasquatch has been reported to be completely covered with black, brown, red-brown, grey, white or silver tipped hair. Black, brown and red-brown coloring is most common with this animal and lighter coloring is usually reported in extreme northern areas of the globe such as Siberia and the Canadian Yukon and Northwest Territories.

The height of the animal has been described as being between eight and fifteen feet tall, the tallest apparently residing in the Alberta area. It is very bulky, weighing anywhere above 200 pounds and very muscular in the arms and legs. Occasionally one is reported to be very thin with a small stature which could be a result of dietary deficiency, age or deformity.

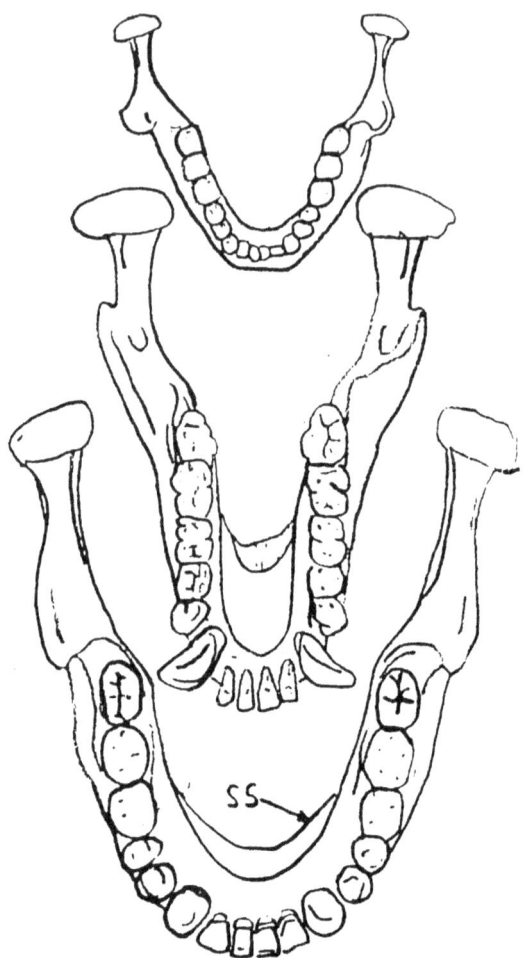

If you compare the lower jaw of the Gigantopithecus (bottom) with the lower jaw of an adult man (top) and the lower jaw of the gorilla (centre), you can see that the Gigantopithecus was huge. Also the spread of the jaw is more like that of the human than the gorilla, whose jaw is much narrower. This argues, in theory at least, that the Gigantopithecus was an erect bipedal hominid (walked upright).

A Sasquatch face is somewhat like that of a gorilla or a very old weathered man. Its mouth is large with big square teeth and occasionally the eye-teeth, when seen are said to be slightly larger that the rest.

The jaw line could Possibly be matched to the Gigantopithecus "Kwangsi Giant" (see diagram on page 7-2), which is extremely close in shape to a Caucasian adult male though more than twice its size.

According to compiled reports (which have been viewed as legitimate), the Sasquatch does not appear to be dangerous or harmful to the human race. In Alberta, the animal has appeared to be either unaffected or fearful of the sight of humans. I know of no account since the turn of the century where the animal was reported to intentionally harm a person.

Older stories from the 19th Century as well as some of the tribal Indian legends convey that the creature is to be avoided. There have been reports of Sasquatch's bluff-charging much like the mountain gorilla is known to do. In these and related incidents the animal veered off before it reached the witness. Assumably, the persons involved in a bluff-charge aged a few decades in the few seconds the incident lasted.

Aside from the above stories, however, the animal remains a possibly shy and even reclusive creature. Some researchers have catalogued it as nocturnal (active only at night).

Habits of the creature, according to compiled reported incidents draw a very crisp picture of its natural instinctive personality. The Sasquatch has been known to scavenge food left on picnic tables overnight yet seem to elude areas lit up by campfires. Some stories recall the animal peering inside tents with the flaps left open. The creature has also been known to shake campers, approach houses and private property, peer into windows and run beside cars (to get a better look at the funny hairless creature behind the wheel). Quite obviously, the Sasquatch has a curiosity that may someday get him into a load of trouble!

The animal seems to have no fear of water and has been reported wading and splashing in rivers and lakes. Taking into account that they can walk faster than a man can run (Olympic runners not included), it is evident that the energetic animal will not be caught until it is good and ready.

Let me, for a moment, take you on an imaginary tour with a Sasquatch, through the foothills of Alberta on an early spring evening. For amusement, we shall call the animal "Stretch".

It is 8:30 p.m., Sunday evening and Stretch has just awakened from a full night's sleep. He moves away from the nest of branches and the pack of family that still lays slumbering on the cave floor, and heads towards the dim light of sunset, creeping through the rocks by the west side of the den. Stretch stinks. Stretch is hungry.

After leaving the cave, he moves through the dense brush with ease stepping from time to time on small bits of broken rock. He approaches the mountain spring and stands beneath the fall, letting the water whip down across his fur and cool his tired body. Next to the spring are four large berry bushes with ripe, red, juicy berries. Stretch leans over to gather a few bushes full (told you he was hungry) then makes his way up an embankment towards a newly hewn cut line.

On either side of the path are animal tracks that Stretch doesn't recognize. They do not begin or end but run up both sides, always parallel. Out of Stretch's instincts dictate him to go farther and discover this new animal in the area. He heads up the cut line, unknown to him, towards the campsite of city-weary outdoorsmen.

The closer Stretch gets to the site of the people the more his nose bothers him. The smell is not familiar and he is increasingly uncomfortable with the sounds coming from just up ahead.

Finally, Stretch sees it! Not one animal but a whole herd! Peering through the bushes, into the clearing, he notices that their paws aren't anything like the tracks they made. Perhaps Stretch is wondering if these animals walk on their heads? Suddenly, one of the animals points its paw in his direction and makes a shrill high scream (mating call maybe?). Stretch lingers for a few moments observing the mass confusion in the clearing then, with an instinct of fear, runs back into the bush, avoiding the track line and steering towards the cave.

Stretch is an animal, so he didn't read the papers the next day nor was he aware of the dissatisfaction of the people with the fact that he didn't stick around to get his picture taken.

Although this is hardly a realistic tale, some of the characteristics of Stretch are similar to believed characteristics of the actual Sasquatch.

The animal is assumed, like other animals of the wild, to have a keen sense of sight, smell and hearing. It is also believed that the animal is vegetarian, although it has been reported as seen eating small rodents. Finally, it is believed that the animal is either extremely curious of humans or genuinely afraid of them.

The idea that the animal is nocturnal stems from the quantity of reported sightings which have taken place in the evening and at night. Those who have reported Sasquatches as being in front of their cars at night state that the animal's eyes shine like that of a cat's shine looking into powerful light. This lends authenticity to the belief that the the Sasquatch has superb night vision.

These, however, are yet theories gathered from eyewitness accounts of an animal who, it is widely believed, may not exist. If the animal is proven to exist, there must be further evidence either uncovered or unintentionally misrelated.

The fossilized remains of a lower jaw belonging to a huge Asian ape, known as the Gigantopithecus and believed to have resided in the area of China and India, may place the Sasquatch into the fossil record. The most interesting thing about this lower jaw is its width, which, when compared to that of a gorilla is considerably wider. The gorilla, whose lower jaw is very narrow, spends most of its time on all fours and its head does not sit on top of its body. The Gigantopithecus, however, has a wide lower jaw very much like that of a human. This has lead researchers to conclude that the Gigantopithecus is an erect bipedal animal (one which walks upright) with a neck. Also judging by the size of the jaw, the animal was incredibly huge.

It is important to note that reports of the Sasquatch describe it as being an upright animal with shoulders and a small neck. Could this giant Asian ape be an ancestor to the Sasquatch? One theory is that the Gigantopithecus moved over the land bridged into Alaska from Siberia at the time the Indians began to inhabit North America. The animals left in China somehow became extinct and those who moved over to North America became known as the Sasquatch. This theory would lend credence to the sighting of similar creatures across Russia and, in fact, various other regions of the world. Unfortunately, aside from the reported presence of more human-like, hair covered creatures (Alma's) north of China, we have no direct evidence that the Gigantopithecus every travelled.

Yet another piece of evidence which could possibly fit into the study of the Sasquatch is a discovery by Dr. Jerold Lowenstein of the University of California in San Francisco.

Dr. Lowenstein has developed a process using immune reactions to identify proteins, enabling him to determine the possible source of feces, i.e: bear, primate etc. In 1987 Bob Titmus of Harrison Hot Springs, B.C. submitted a hair sample collected from trees. Dr. Lowenstein was able to identify the hair as coming from a primate, yet he could not identify whether it had come from a human, chimpanzee or gorilla.

"He therefore assumed it was human, but in fact it was nothing like human hair, since it had wool hairs as well as guard hairs and the guard hairs were pointed, not cut off at the outer end. Unfortunately he had ground up the whole sample, otherwise it would have been simple to check his three possibilities with a microscopic comparison. I am quite sure all three would be eliminated, as I doubt that chimps or gorillas would have wool hairs either - this is a point not covered in any reference ... except that lowland gorillas definitely have no wool hairs." (John Green, Letter September 1988)

Most definitely a higher primate was leaving hairs on the trees in California, a primate who could not be conclusively identified by advanced methods of identification.

CHAPTER EIGHT
"A Hunter's BIG Bear?"

On the night of October 20, 1987 an unidentified caller reported that he and a friend had seen what they believed might be a Sasquatch in April of the same year.

According to the man on the other end of the line, he and a friend were on a hike along a cutline a few miles northwest of Rocky Mountain House, as the sun was setting. Both men were apparently hunters and well versed in the identification of animals. Then, out from the tree line stepped a 7 or 8 foot tall, dark hair-covered creature standing on its hind legs.

The two stopped dead in their tracks and stared straight ahead getting a full view of the strange and unidentified, to them, animal. The animal also saw them and bolted back into the tree line and out of sight. According to the two witnesses, the animal ran back on its hinds.

These two men were not the only hunters to have ever sighted a similar creature. Hunters are a unique brand of people. A good hunter is disciplined in animal identification and tracking. Because Alberta's hunting laws are stringent, men tracking down game must be careful to identify the target as an animal he is licensed to shoot. Not just the animal must be identified, but also the specific breed - if the wrong breed is shot in the wrong season a hunter can face hefty fines. However, I have on file a good number of reported sightings from hunters - not the type of report to dismiss as a case of mistaken identity.

The earliest Alberta Sasquatch sighting in my files took place in late summer of 1954 north of Blairmore and Coleman by Mr. Lidio Orlando, a hunter who lived just north of Edmonton.

Mr. Orlando wrote me a letter to tell of his sighting. Following is an excerpt of that letter.

"...It was open hill country with mixed spruce, pine and some poplar. The highwood range was to the east and the Rockies to the west. I was out hunting rabbits with a .22 calibre Winchester and my dog King. About 4:00 in the afternoon I was sitting on high rock ridge looking east, when I spotted a tall black creature walking swiftly towards me. I just sat there watching, thinking it was a man at first, but the closer it came the more unman-like it became. Walking erect, it covered ground at an unbelievable rate. No human could possibly run that fast, let alone walk. When I first saw the creature, it was approximately 400 yards away. It walked directly towards me to a point about 180 yards east of me, then turned right and headed northwest up a ravine until I lost sight of it.

... It walked directly under a large tree, with a large overhanging branch about 16 feet off the ground (and) ... cleared it by about 2 or 3 feet. ...the most striking thing I can remember is the size, and the gait which was unbelievably fast."

What was it that Lidio Orlando saw in the summer of 1954? Could it have been a bear? Bears do not walk on their hind legs. Could it have been someone playing a joke on Lidio? If it were a joke, the man playing it must have been very tall and extremely energetic, beyond any man known in history!

At 4:00 p.m., June, 1974 in Bragg Creek Provincial Park Carl Melnyk a hunter saw what he believes might have been a Sasquatch, while spending the afternoon with friends at the campground.

Carl was sitting on the bank of the Elbow River. Approximately 100 metres away, just inside the tree line, something was moving. The movement caught his attention and he glanced over. A large dark brown or black body covered in animal hair came into view. Naturally Carl thought the body mass to belong to a moose or elk.

Then, Carl's eyes grew wide and he strained to get a better look at what was in the trees; why - because it was walking upright, on two legs, just like a person! He watched in amazement for close to 15 seconds before the animal slipped out of sight. Carl quickly went to the area where he had seen the creature and searched for tracks but none were found. Later he told his friends about the incident - they laughed.

Carl's curiosity was not quenched, however, until 13 years later. When he returned to his apartment in Calgary Carl saw my ad and called. He wanted more information on the Sasquatch - precisely what I had asked for in the ad.

According to Carl the animal he saw was at least 600 pounds, big, bulky and husky with long thin arms. What convinced him that it was not just a large man?...

"After a few seconds I noticed that it was moving on two legs and that is when I realized that it was not a normal animal, at least not a common one. So I just stared at it for about 15 seconds and then it took off in a big hurry ... ripping through the trees like there was nothing there ... it was the speed that convinced me this was not a man!"

On the 10th of July, 1988 three teenaged boys and an experienced hunter fled a campsite, panic-stricken. Later that evening, the R.C.M.P. and the local Warden's office were contacted and given what could have been a "bear-scare" story.

A close look at the report, though, led me to believe that an animal other than a bear caused the panic!

A Lethbridge man had taken three young boys camping and backpacking. After taking possession of a cabin by Upper Twin Lake in Waterton Lakes National Park the foursome headed out for an afternoon of fishing and backpacking. Still within hearing distance of the cabin, they settled by the lake with their fishing rods, bated the hooks and dropped the lines - all hoping for a prize catch. Suddenly a commotion coming from the direction of their cabin broke the afternoon lull. A high pitched scream echoed through the trees - a scream that none present had ever heard.

As mentioned earlier, the leader of the troop was a hunter who had heard wild animal calls many times before but this was not an animal call he could place.

When the boys returned to the cabin their peaceful camp had been disturbed. Something, someone was prowling through equipment and personal belongings. Fearfully, they packed up immediately, and left the area, not checking for tracks or signs of who or what was driving them away.

Sometime later I was contacted by these people. At that time I had in my possession a cassette taping from Fouke, Arkansas, of what is believed to be a Sasquatch call. During our interview the tape was played. All four agreed that the sound was exactly the same as the high pitched scream coming from the direction of the cabin.

Most possibly the scream could have come from a human intruder using scare tactics in order to rob the site or even drive the boys from the cabin. It could have come from a wild animal who was hurt or dying. Why not? Nothing was stolen from the sight - just disturbed; as for the animal theory it doesn't work for the same reason - a wounded animal would hardly just disturb a campsite then quickly disappear from the area.

It would not be presumptuous to assume that if the Sasquatch does exist it has a distinct call or cry. Many reported incidents of strange "screams in the night" unidentifiable as human, bear or any known animal have come from areas frequented by actual Sasquatch sightings. In 1978 a family reported a very similar incident.

The Simms family (alias) were off to spend a day at Jackfish Lake, north of Highway 11 between Nordegg and Rocky Mountain House. Mr. Simms brought the family car to a stop. The family piled out of the vehicle and began unloading the car, setting up day-camp. The area was dense with bush and trees. Birds were peacefully chirping, small creatures were scurrying across the landscape. then, in a split second the calm was broken.

The birds ceased singing and took flight from the area, small critters scampered to safety and one incredibly long, eerie scream rang out through the trees, across the lake filling the sudden silence. Mr. Simms blood ran cold - he had never heard anything like it. The whole family stood staring toward the bush. Something big moved quickly deep into the tree line.

The Simms' threw their belongings back inside, scrambled into the vehicle and drove off. Not one of them would brave investigating.

Mr. Simms has never forgotten the sound of the scream.

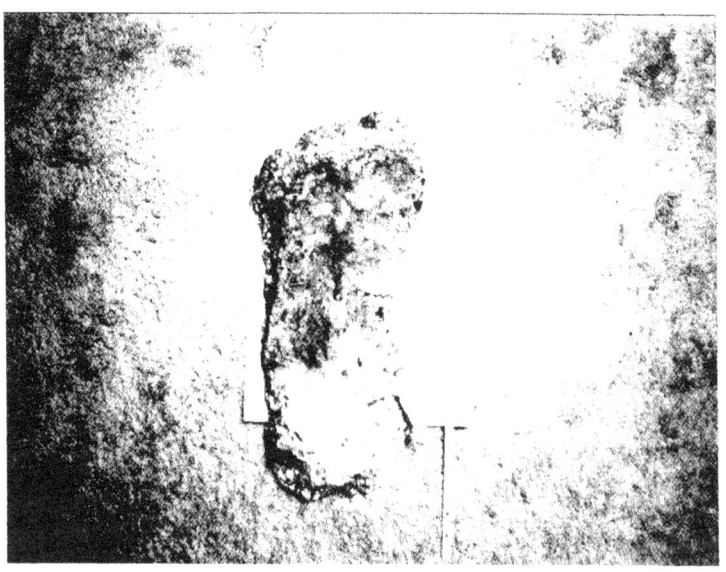

Cast of footprint found by author on August 1986. Chilliwack River.

Mr. Donald B. Kenny was a single shot away from solving the mystery of the Sasquatch once for all.

Donald Kenny didn't answer an ad in the paper, he made no report to the R.C.M.P. or Warden's office but what he encountered, what he had seen with his own eyes during a 1975 hunting trip warranted telling someone —

someone who would take his story seriously. Donald Kenny wrote to John Green, a well known author and authority on the Sasquatch search.

John Green forwarded a copy of that letter to me. I contacted Mr. Kenny and requested an interview.

On November 7, 1975 Donald was moose hunting fourteen miles west of Sundre, Alberta. He fired a shot at the Sasquatch and missed. I asked him to explain exactly what happened.:

"It was roughly 8:30, 8:45. I had stopped my car along this forest road because I'd seen a bull moose in the swamp. It was almost right over on the north side of this swamp. I got out of my car, loaded my rifle and took several steps away from my car into the ditch. The moose was standing broadside to me, I aimed and shot. The moose did fall down. As it fell, its four legs dangling in the air, I tryed to reload the rifle again but it was jammed. While I was busy fidgeting I looked up and saw this moose was getting back up on its feet. It proceeded to move rapidly away from me, heading up-hill onto the cow trail. At the same moment that the moose reached the trail a Sasquatch walked onto the same trail. Both animals then disappeared from sight.

I went to the place where the moose had fallen to check for blood because, believe it or not, my primary concern was to get the moose. When I checked and found only a little bit of blood I figured I'd better go after the moose and see if I could get it.

I went up the hill toward the trail in the same direction as the moose and the Sasquatch. After going about 200 yards the tracks became scarce. The area was above the snow and the ground was covered with a spongy moss. I couldn't track the moose very well.

I circled the area in a wide circle, then an even wider circle. I had just finished the second circle and was about to start a third when I had come into a small clearing. The sun was coming in from the east side, I looked east and could not see any tracks, I looked south and still couldn't see tracks.

Then, for some unknown reason (hunter's intuition) I just knew that I was being watched. Slowly turning my head to the left my eyes met the eyes of the Sasquatch. It was only 90 feet away (approximately). Then I felt fear. Taking the safety off my rifle I started to bring it around and the Sasquatch took off.

I thought it was going in a northeast direction, so I aimed for a spot between two pine trees and fired. After this I made another circle to see if I could see it again because I did not know if it had been hit. The circle brought no results - I did not see the animal again."

Mr. Kenny also gave a description of the animal. He claimed the animal's facial features were comparable to that of a human's and covered in reddish brown hair. It walked exclusively on two legs, not once going down on all fours and its arms hung below the waist.

The animal he alleged to have seen that day made no sound other than sounds caused by natural movement. It did not bear its teeth or show signs of aggression.

Mr. Kenny saw what he believes was a Sasquatch. He saw it in broad daylight twice. He saw it clearly and he fired his rifle with the purpose of hitting it. Had Donald Kenny actually gunned down the animal and brought its corpse (or injured body) into civilization the mystery would be over.

Two more sketches made by
'An Observer' Medicine Hat,
December 1973.

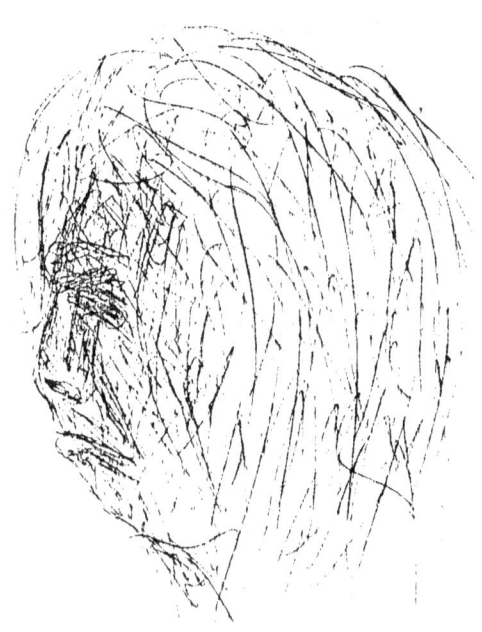

CHAPTER NINE
"The Clearwater River"

From the town of Caroline to the Rocky Mountains lies the Clearwater river area and for the past 15 years it has been a buzz of Sasquatch activity. Unlike most camping and wildlife areas in Alberta, the Clearwater River area is widely used for recreation, camping and wildlife exploring.

Amidst the wild untouched landscape, mazes of dirt roads, cut lines, and occupied campgrounds twist around the area. A drive along the banks would afford the pleasure of seeing wildlife - elk, deer, black bear and sometimes grizzly intermingled with the human trappings of partial civilization. Phyllis Lake Campground, Tay River, Seven Mile Recreation, Elk Creek, Cutoff Creek, Equestrian Campground, Peppers Lake and Ran Falls are widely used by people - yet the animals remain in the area. This could be the reason for such frequent sightings of the Sasquatch; not that there are more Sasquatches in Clearwater River, but that there are more people to see them.

Just outside of the Phyllis Lake Campground on the banks of the River on a pleasant September afternoon in 1976 two men were hunting grouse. Gary Schmidt of Alberta and Peter Garyall of Alabama, U.S.A. walked parallel, Gary on the south bank and Peter on the north bank. Their hunting efforts that afternoon were futile.

Peter walked along until he came to a bend which gave him a clear view of the river. Something piqued his curiosity. A large form about 200 yards down the river squatting by the water, splashing it over its body. "Hey, have you got apes up in the mountains?" he yelled over to Gary.

Gary, caught off guard by the question raced over to the north bank by the bend to get a look at what Peter was calling an ape. He looked down 200 yards and stood watching, stunned. There along the waters edge, not a block distance away was a creature neither man had seen before - it almost looked like an ape.

The creature was about 9 feet tall, covered in black hair, its head was sitting on top of its shoulders, not jutting out from just above the neckline. Its arms were incredibly long and its face had the appearance of a cross between a human and a dog.

The heavily wooded landscape just north and south of Highway 11 is where a high number of Alberta's Sasquatch sightings seem to occur.

For nearly five minutes the animal washed itself and the two men looked down upon it. The shallow, rushing river drowned out any sounds and neither Gary or Peter could hear anything coming from the animal.

Then it looked up towards the bend, spotted its observers and hurried off into the tree line and out of sight. Neither Gary or Peter reported the incident to authorities.

Sometime later, Vladimir Markotic, professor of Archaeology at the University of Calgary who has been researching the Sasquatch mystery since the early 60's landed a spot as a guest speaker on a radio talk show. When the talk line was opened, Gary Schmidt called Professor Markotic and reported the incident.

The public's curiosity cat sprang into action. A few months later I was invited as a guest on Calgary's radio talk show "The Home Stretch" to discuss the Sasquatch. When the talk line was opened Gary Schmidt was on the other line wanting to know more about these creatures. The name was familiar, Vladimir had passed on this man's name and phone number to me and I had called Gary to hear his story.

Gary wanted to know more about the Sasquatch, I wanted to know more about Gary's sighting. I requested an interview. He was not too enthusiastic about being interviewed, however. Gary had been on the receiving end of jokes and jeering because of the sighting. He was, of course, not alone. I explained that many people who previously came forward with a story of a sighting required more courage to face the suspecting and critical public than to face an angry Sasquatch head on. Gary decided then to use his courage and agreed to meet with me.

Q.   How much of the creature did you see and for how long?

A.   *I saw the whole thing for about five minutes. I was no more than about one city block from it.*

Q.   Could you describe what took place?

A.   *"I was hunting rough grouse with a friend of mine along the Clearwater River near Caroline. We were walking along when he yelled up to me saying 'you got apes up here in the mountains'? I went running down and I saw this big animal in the river throwing water on top itself.*

It was splashing itself with water. I ran down toward it a little ways and my friend says 'are you crazy? Such a big animal!'. After about five minutes it spotted us and just got up and took off."

After the interview we sat and talked for awhile. Gary told me how he had been made fun of by his friends and some of his family. Later we shook hands and as I walked out the door his last words were "Mr. Steenburg, you know now I wish I never saw that thing... Nobody believes me and so I just don't talk about it anymore."

Another Clearwater River incident took place in the first week of August 1984. The man involved does not wish to be identified publicly so I call him Bud.

Author searching for footprints, Clearwater River.

Bud was camping with his wife at the Seven Mile Recreation Area. As dusk was falling He headed out for a walk to stretch his legs. Bud walked down a cut line lined with trees between 9 and 20 feet tall for about 4 or 5 city blocks.

Tay River Campground, 1979. Vladmir who is over 6' stands at spot Sasquatch reportedly crossed the road.

Bud developed a strange sensation that he was being watched. He looked around and just inside the tree line He saw the head and shoulders of a creature that must have been about 12 feet tall. It moved for 3 to 5 feet then disappeared into heavier bush. Bud could hear its footsteps. The animal sounded very heavy and slow its every step sounding a loud thud. Fearing for his safety, he turned and headed, slowly, back toward the campsite looking over his shoulder every couple of seconds, afraid that the creature, whatever it was, might come after him.

Bud waited until the next morning to tell his wife. He didn't want her to be afraid throughout the night.

In 1979 in the Tay River campground Mrs. Diane Menzal and her son were walking along the highway near the river crossing. They heard a noise behind them and turned to see a large very tall hair-covered creature cross the highway, look at them, and walk up a hill on the other side of the road and into the trees.

The creature did not break into a run, it just walked like a man would walk - remaining on its hind legs the whole time. Afraid of ridicule, she kept the story to herself until hearing of a forestry worker in the area who almost hit a Sasquatch with his vehicle. Diane came forward with her story and reported the incident.

Some months later Diane's son called me to report the incident and tell his story as well. His story collaborated Diane's except for his failure to mention the creature turning to look at them before it walked off.

At the date of writing, the forestry worker has not been traced to further confirm the sighting.

CHAPTER TEN
"Additional Sightings"

A man from Drayton Valley, Alberta wrote John Green with a reported sighting in the Lake Ribbon Creek area in April, 1969.

> It was around noon time that the three men built a campfire to heat up and eat some food they were carrying with them. They sat and ate and talked for about 30 minutes when one of the men said "Hey what's that?". His two companions looked to where he pointed and saw about 100 yards away from them what they later described as a gorilla sitting on his haunches watching them. Two of the men wanted to run back to their car but the third man whispered to them, "wait a minute". Apparently, the animal just sat there looking at the men for about five minutes, then it stood up. The creature then made a chattering noise with its teeth and at the same time moved its arms in an up and down motion. Later, after the creature had moved off, the men went to the spot where it had been sitting, but they did not find any footprints nor any other sign of the creature. Later, all the men agreed that the strange animal stood between seven and eight feet tall.

The man drew a sketch of the animal and included it with the letter. According to the sketch, the animal may have been female as it had long droopy breasts.

This sighting was in the same year as the Edmonton Journal report of Chief Joe Smallboys Indian band sightings (Chapter 1).

In August of 1968 Mr. Gerald Martin claimed that he and his family watched what might have been a Sasquatch walking along a ridge east of Highway 93, opposite the Columbia Ice Fields. He described the creature as a large, black upright figure walking along a ridge quite a distance away. The family agreed that it was too large and walking too fast to have been a man.

Guy Phillips, a researcher in Winnipeg, Manitoba wrote me about a report from the Lake Louise area in December 21, 1984. Guy has been looking into reports for many years and when he hears of something in the Alberta area, he will usually correspond the details to me.

Guy's letter regarding this incident follows:

"As I mentioned on the phone, I have some info on a sighting out your way from 1984. The details are very sketchy but you are welcome to them. I wrote little down at the time at it wasn't local and I relayed it verbally the same day to someone who passed it on to (Rene) Dahinden.

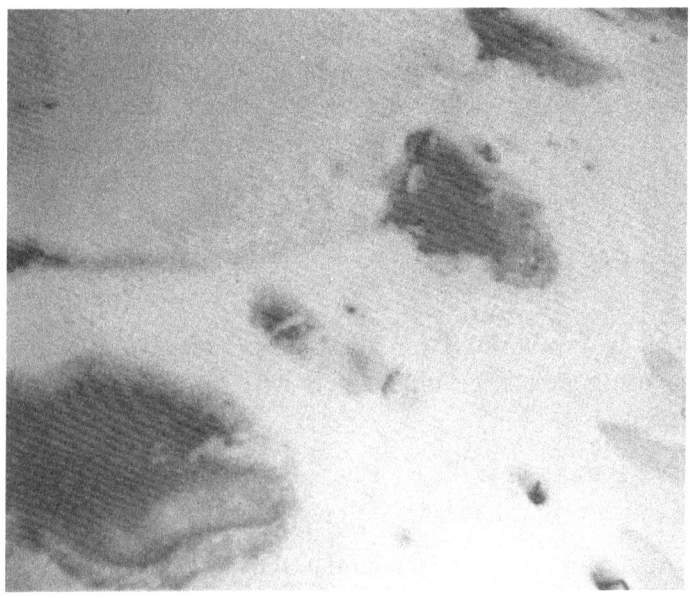

One of the photos taken by Sharon Smith (not her neal name) of the tracks she and her husband thought might be Sasquatch tracks. But I am sure they are just footprints melted out.

It took place (between) December 18 and 21, 1984 (she wasn't sure of the exact date). Some university students from Winnipeg were on a ski vacation. At one point they were driving just outside of Lake Louise when they saw an animal climbing up a mountain. It was eight to nine feet tall, a brown (dark) or black color. They (3 or 4 in a car?) stopped the car and watched it for 5 or 10 minutes. It never went on all fours and continued up the mountain side. They later returned to the chalet they were staying at and were overheard talking about their sighting by a man there. He told them he had been hiking in the same area several months before and had found and photographed some 17 inch tracks he had seen in the mud. I spoke to the girl who saw it, in August of 1985 and I think she believed she saw something that was not a bear. This incident took place at dusk, however, and using local reference points she put the distance at approximately one or two blocks. She had no idea who the man at the chalet was. They were not drinking at the time of the sighting."

Unfortunately, the report did not make it to Alberta until just recently. Therefore, the story can not be properly investigated and the man who allegedly photographed the tracks has not been identified. Also, no reports of these pictures have reached my files.

On the night of March 21, 1988 Dan and Julia (alias) reported to me their sighting of what they believe was the Sasquatch.

In May, 1983 on the Stoney Indian Reserve along Highway 1A crossing Oldfort Creek, Dan and Julia were driving over the bridge when they spotted a dark brown human-like creature step onto the road. The couple immediately identified the creature as a Sasquatch.

The creature turned to look at the oncoming car then bolted across the road, jumped a four foot fence and ran into the trees. Dan wanted to stop the car and chase after it but Julia, who was pregnant at the time, became very frightened and begged him not to go after it.

"Just keep driving" she pleaded, and Dan did just that.

The couple saw the creature at a very close proximity. It was covered with hair except for parts of the face which were covered with black skin. Dan described the eyes as large and deep set with the forehead jutting out. He claims that it stood between seven and eight feet tall with large and very muscular arms. Both the husband and wife agreed that it was remarkably fast and bolted over the fence with very little effort. When they drove past the spot on the road where the creature had jumped over the fence and into the trees, they could see the trees and brush moving as it ran deeper and deeper into the woods.

Track from 1986 siting, Chilliwack River, B.C.

Track from 1986 siting,
Chilliwack River, B.C.

When Dan and Julia first reported the incident to me, they were both willing to have their names and place of residence published but later withdrew because they "wanted to put the incident behind them and go on".

Dan is a hunter however, and expressed a desire to hunt the creature.

Not all Sasquatch incidents reported are authentic. Many actually check out to be false. Occasionally I receive reports by people who sincerely

The 4' fence which the Sasquatch cleared with no effort at all, according to Dan and Julia.

believe they have found something which relates to the Sasquatch but the evidence is merely a case of mistaken identity.

On the night of January 26, 1987 a woman called to report a track find in the snow on a frozen swamp area not far from Rocky Mountain House, discovered in 1986. Her and her husband photographed the strange tracks and wanted to present them to me for verification.

There were six photos in all and they did indeed show large impressions almost walking in a straight line across the ice. The size of the prints were huge when compared with Sharon's 7 1/2 size running shoe along side in the photograph. The couple did not measure the size, however, or the stride. Upon close examination I concluded that the prints were not that of a Sasquatch but of a human — the prints merely melted out in warm weather.

The outline of a boot print inside of each track could be seen. That was probably the original size of the print but the warmth and the rays of the sun on the ice seemed to have melted them causing the larger impression that convinced the couple that the tracks may have belonged to a Sasquatch. Still a further clue in examining the photos was the size of the gait or the distance between the tracks. If the tracks were their original size, whatever made them was walking heel to toe at the time.

The courage of this couple to come forward and the intelligence to take photos was encouraging though. I thanked them for calling and expressed my desire to have more people come forward with possible track findings.

The bridge over the Oldfort Creek on which Jim and Julia encountered a Sasquatch in early May 1983.

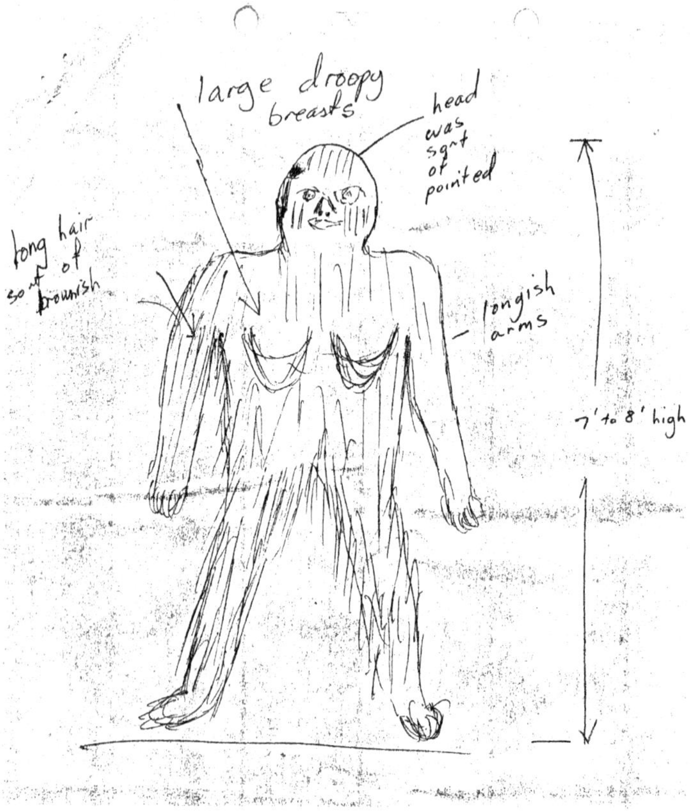

large droopy breasts

head was sort of pointed

long hair sort of brownish

longish arms

7' to 8' high

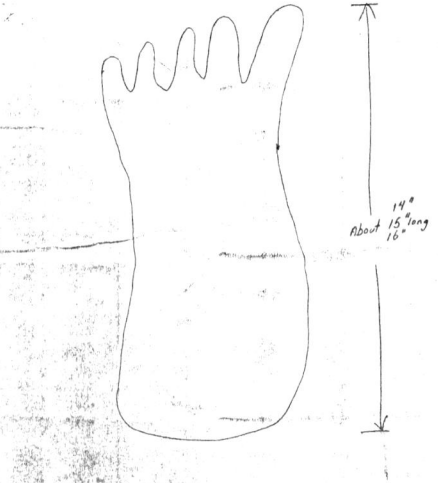

Top: Witness sketch of figure seen near Ribbon Creek April of 1969, (Chapter 10, page# 69, top)

Right: Footprint sketch from same incident.

About 15" long
14"
16"

CONCLUSION
"To Shoot Or Not To Shoot"

There have been many self-proclaimed experts in the field of Sasquatch research. Some have thought to know the migration routes, caves which they occupy, rivers they seem to favor but in the opinion of this writer, there are, as yet, no experts.

Many excellent researchers are presently working towards finding a Sasquatch and proving its existence (some of which are quoted and give attribute in this book). Until, however, one of these animals is brought into captivity and examined closely and thoroughly, there will be no experts.

A popular debate among those who believe in the existence of the animal is whether or not to shoot a Sasquatch in order to prove it as a legitimate animal or breed of animal. Now, before every naturalist and environmentalist raises arms against me and threaten flogging, I will make it clear that shooting one of these creatures is not an idea I personally relish. In fact, before 1980 I was dead set against it always hoping that myself or someone else would be lucky enough to come across the remains of one who died of natural causes.

Unfortunately, dead animal flesh does not last long in the wild. Scavenger animals and birds will most often destroy a corpse before man lays his eyes upon it. Also, it is a rare occurrence to find a whole skeleton in the wild in one piece.

When a handful of men began seriously investigating the question of this animal they met with closed minds in the scientific community whose overwhelming attitude in the late 50's and all throughout the 60's was that if such a thing existed, they would surely be aware of it. Although science has witnessed some advancement in this area, the attitude remains very much the same. Most scientists claim neutrality on the subject of the Sasquatch.

"I will keep an open mind if you can present me the body of one of these creatures" closely reflects the current thinking. Very few scientists are willing to conclusively examine current evidence although many will suggest that they are convinced that there is truly something out there.

Professor Grover Krantz has studied the anatomy suggested by the footprints. He has concluded that in at least some cases the tracks in question could not have been faked. In fact, a set of footprints that Professor Krantz studied, found in Umatilla National Forest on the border between Washington and Oregon states in 1982, were clear enough to show skin marks known as dermal ridges (like finger prints on the bottom of the feet). Most finger print experts who have examined these tracks concluded that they were real. My

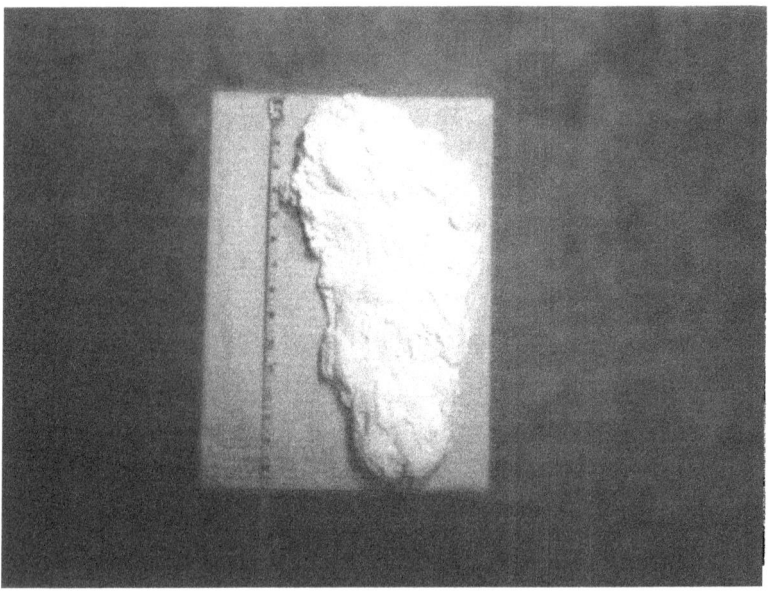

Cast of Alleged Dermal Ridge Footprints, Umatilla National Forest, 1982 (left foot).

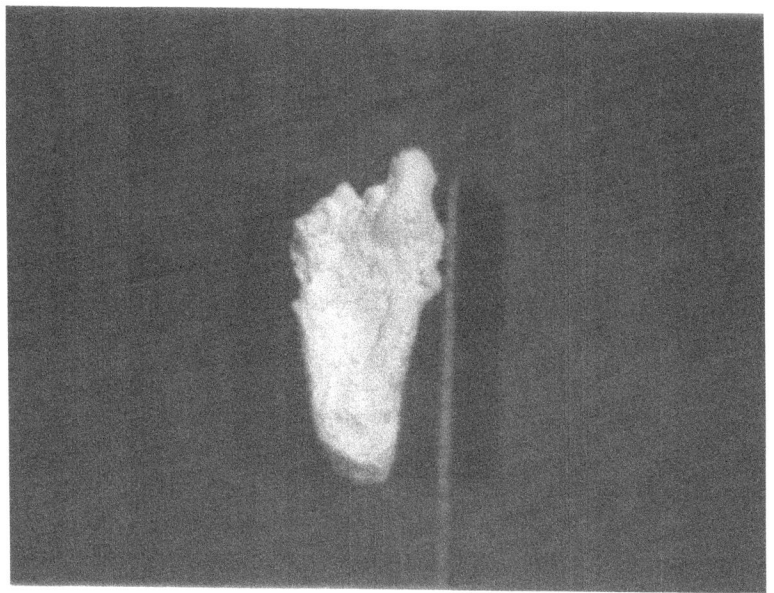

Cast of Alleged Dermal Ridge Footprints, Umatilla National Forest, 1982 (right foot).

natural reaction was that this would lead to a serious confirmation by the scientific community that the Sasquatch is a definite possibility. It is unfortunate however that I must now share my doubts. Because the individual who first found the tracks has openly admitted to faking prints in the past, we cannot be one hundred percent sure that they had no hand in faking the dermal ridges. It has also come to my attention that given the right soil conditions etc. it is quite possible to fake dermal ridges.

Although the dermal ridges turned a few heads, little changed in the way of attitudes towards further research. Realistically speaking, the objectivity of science would demand that this information merely be filed and held for comparison with further evidence. It is this realization that has turned me around in my "shooting" question.

I am persuaded that at least one must be shot in order to prove its existence and bring it into the historical, environmental, and scientific realm. Without proof of its existence the animal cannot even be properly protected from possible extinction.

One notably different opinion on the "shoot or not shoot" question is that of my friend Professor Vladimir Markotic (Archaeology, University of Calgary) who has been avidly involved in researching the Sasquatch question since the early 1960's. Vladimir is unique in that he is the only legitimate scientist I know of who is against the shooting of a Sasquatch to prove its existence. He said to me some time ago "Tom, why should we have to prove anything anyway. If you shoot a Sasquatch and say here is the proof, the scientists who said there was no such thing will claim that they knew of it all along". No doubt Vladimir has a point here. He also said "We should not look at the subject so black and white. If the Sasquatch is real, then it is an important scientific discovery, if it is not real then it has to be looked into anyway because it is an important piece of folklore."

Vladimir has given new food for thought lure. I must admit that the only thing I was thinking of was whether or not the Sasquatch was real. So is the Sasquatch real? Is there a large bipedal animal running around in our mountains, forests and bushes?

This is for you to decide. My personal opinion is that the animal does exist. To date, there has been a total of 3000 reported incidents and the number grows by the day. Many of these reports were made by reputable people, who do not personally know others who have made the same claim and told of the same encounter.

There is also the Patterson film, the alleged dermal ridges, many tracks and clear prints, eyewitness testimonies, historical sightings from years ago, Indian Legends which tell similar stories of a similar creature. I am, however, willing to be convinced that there is nothing out there if the evidence eventually convinces me.

The question of the Sasquatch will rage on as one of the most debated subjects of this century until one is either shot or brought into captivity.

Food for thought: An unsolved mystery; a hoax; or an Indian legend that lives on.

Author in his study, Water Valley, Alberta, 1990. This work had just been published the first time.

--UPDATES SINCE ORIGINAL 1990 PUBLICATION--

Investigations after 1989 for The Sasquatch In Alberta

Since the decision to republish this work of 30 years ago, the publishers thought it might be a good idea to update the work a little bit. After all, 1989/1990 was a long time ago and also, my investigations into this ongoing mystery have continued. In September of 2002, I moved away from the Wild Rose Province to the Fraser Valley of British Columbia. The result, of course, was loosing track of on-going reports from the east side of the Rocky Mountains. But others have taken up the call, and on occasion, I do hear about Alberta Sasquatch incidents. I also did spend another 12 years doing research in Alberta after this book was originally published and authored two other titles in which some Alberta incidents were included. What follows now is an overview of possible Sasquatch encounters that I did investigate after the original publication of this work. I hope the reader will find them interesting and keep in mind that the existence of this creature is a possibility in Canada's Wild Rose province.

August 4, 1996.

Mr. B. Adlington reported to me that while he was camping in a forest clearing on the 940 Trunk Road just south of the Livingston Falls Recreation Area on August 4th, 1996, he had an encounter with what could possibly be a young or juvenile Sasquatch. He had pulled his camper truck into the area the night before in preparation for an upcoming dirt bike event, which over the next number of days, would bring many more people to this area setting up camps and just generally turning the whole area into a small temporary community of off-road enthusiasts. But for now, he had the whole place to himself. At approx 9: 30 in the morning he was walking with his dog enjoying the early morning crispness, a cup of coffee in hand. Some other campers had arrived just a few hours before. He was about to head back to his own camp when he heard a wailing cry which kind of reminded him of the sound a bear cub makes when in distress. His own words are as follows to one of my questions when I interviewed him:

"We were camped below this meadow. So I climbed this steep bank, when I got to the top of the bank I was level with the meadow. I had took a few steps forward when I heard the noise it was making. I stopped in my tracks, and the dog stopped and we both focused forward. I looked up and I could see him against the trees on the far side of the meadow, and he was traveling from where I was standing, right to left, (North to South) and while he was running I was watching and listening. I was hoping he would stay out in the meadow so I could have a longer look or even come closer but after about 15 seconds, he disappeared into the trees. I watched for awhile but never saw any other signs."

The witness went on to describe that at a distance of about 900 - 1000 yards, the figure appeared to be covered in black hair, head to foot. It moved on two legs the entire brief time it was in view. He thought it was between 5 to 6 feet tall; it was not slim; and estimated between 200 - 300 pounds. Too far distant to see any facial detail and no idea as to gender, he saw the thing for about 15 seconds as it moved into the trees, out of sight. The sound the thing made, he described as similar to the noise a bear cub makes when in distress. Yet the witness is certain this was not a bear he was looking at. The dog was watching the subject with interest but did not bark or try to pursue. The witness remained in the area as other people arrived and the off-road event went off without further incident. As far as I was able to confirm, no other event attendee reported seeing anything odd during this time. I did interview the witness and went to the location and found nothing in the area to confirm that a Sasquatch could have been there. On the other hand, there was nothing to prevent a Sasquatch from being there either as this part of the Alberta Rockies is known for its wildlife. So, what was it Mr. Adlington saw that morning? Could it have been a bear? Or a person in dark clothing? He does not think so.
At the time, I found no reason to doubt his conclusion.

August 17, 1993

This incident was forward to me by two colleagues and good friends at the time, the late John Green, and the late Ray Crowe. Both men exchanged info with me and each other when we were made aware of events in our areas. In this case, two witnesses whom had a strange encounter on a public hiking trail in Jasper National Park, Alberta. It was approximately 6:00am, on August 17th, 1993. One of the two witnesses wrote out a statement:

"We were on the hiking trail which at this point went along the bank of a stream. Well, at first we hear this strange hooting type noise? It came from about 30 meters off the trail. We wondered what it was, so we went to investigate. Being cautious as possible, we crept closer and saw this large hairy man. It stood about 9 feet tall and was built bigger than a gorilla. It seemed to be looking at something or someone. It turned and looked at us and I saw in its eyes a gleam of intelligence and wild fury. We ran as fast as we could away from the beast and it followed. We ran for about 20 minutes. It seemed that it could have caught us at any moment but it seemed that all it wanted to do was chase us away from what we do not know. When we went back to that location 8 hours later, we saw many large tracks, bigger than the largest shoe. We also noticed there was more than one set of tracks. Unfortunately, it started to rain hard shortly after that and the tracks were lost."

It appears that the witnesses never saw a second creature. Nor did they give any contact information to either Ray Crowe or John Green. Researcher Scott Mc-Nabb, of the BFRO, also looked into this case and managed to talk to one of the witnesses. I too, was able to talk to one of the men over the phone and he told me that some of these tracks were 3 inches down in semi-dry mud. That must have been one large downpour of rain to wipe out tracks imbedded 3 inches in partly dry mud. Neither witness wanted to be made public so I will just refer to his first name, William. Now William came away with the impression that there had to have been two creatures in the area, as they found two sets of footprints. They only saw the one with dark hair from head to foot, approx 9 feet tall and built bigger than a gorilla. Well, I came away thinking they either encountered a Sasquatch or they like to make up stories. Though wanting to keep confidential seems to argue against attention-seeking.

December 12, 1998

A friend and colleague, the late Rene Dahinden, forward to me a letter he got from a Mr. Redmond, who had a friend whom told him he saw a Sasquatch on the road near the town of Grand Prairie, Alberta during the late fall of 1998. Rene thought I might want to follow up on this. He was right. I contacted Mr. Redmond and found out the name of one of the witnesses; one Jack Durant. He lived in Grand Prairie and was surprised to hear from me and wondered how I had found out about it. After some time, he told of the encounter.
Mr. Durant was driving a company truck south on Route 40 about 20 miles south of Grand Prairie at approximately 9:30am, on December 12th 1998. A co-worker was in the truck with him. This man was new to the job and had recently moved to the area from Vancouver, BC. His first name was Stewart. The two had just passed the Windy Andy Road turn off, and were driving into

the Smokey Valley, which contains the Smokey River. Jack saw to his aston-
ishment on the high tree line on the right hand side, (west side) of the road,
a dark brown hair-covered creature which seemed to be deciding whether or
not to proceed with this truck coming towards it. At this point on Route 40,
there is a hill between the road and the tree line. Jack told me that the creature,
after a moment of hesitation, it seemed to stop itself by grabbing a tree with
its left hand, then let go of the tree, all in a fast, yet graceful manner. Quickly
advancing down the hill, jumped the ditch running along the roadside, and
crossed the road in front of the moving truck, which at this point was perhaps
150 to 200 feet from it. It was moving at a very fast pace when it disappeared
in the tree line on the roads east side. The entire sighting was about 30 seconds
in duration. Jack described the creature as ape-like. It stood about 7 feet tall
and walked upright like a man the whole time it was in sight. It was covered
in either very dark brown or black hair and had dark gray skin on the face,
which was also hairy in places. The creature never turned to face them when it
crossed the road but seemed intent on crossing the road as quickly as possible.
Jack said that when he first sighted the creature, it was facing toward the truck
and seemed to hesitate for a moment by stopping itself by grabbing a tree. Jack
did not stop the truck but continued to drive on, both men talking loudly to
each other about what just happened. There was about a foot of snow and the
two men did see a path it made but did not return to have a good look at it. It
was a sunny day, but cold and a little windy. On the return trip after dark, the
trail left by the creature was still there but had been nearly filled in by blowing
snow. Jack also told what most impressed him about the creature was the speed
and grace it displayed when it moved. It never broke into a run but walked very
fast as came down the hill and crossed the road, in a foot of snow, leaping the
ditch on the road's right side. The second witness in this incident declined my
request for an interview.

July, 1995

A report by two hikers came to my attention by fellow researcher Dawn
Harrack, a member of the BFRO (Bigfoot Research Organization). This incident occurred on a popular hiking trail near the town of Jasper, Alberta (Jasper
National Park). The witnesses were hiking up the trail towards the summit and
had stopped to take a break and to shoot a number of photographs of the hot
springs and local mountain scenery.

"A friend and I were hiking on the Suplur Ridge Trail; we came to a place where the trail had fallen away, so we decided to stop for a break. I took out my camera to take a picture of the hot springs below. I happened to look up at the summit and I noticed someone standing up there. I started waving because I thought it was a person, and since it seemed we were the only people on the trail, we'd be walking past them anyways. I even pointed my camera at it, but didn't take a picture due to the distance. The thing didn't wave back, and it was then I really took a look at it. It was about 6:00pm, so there was plenty of light still. What I saw looked to be a dark brown man covered with hair. It stood upright and had arms and legs like a human. After several minutes, it started to run down the face of the mountain; we were below the tree line, but well within sight of the summit. At that point, I said, Holy shit, that guy is nuts! I now thought it was a mountain biker because he descended so quickly. About one and a half minutes at the most, it came straight down without using the switch backs and that's when I knew this thing was not human. I thought of a grizzly or other animal. When it hit the tree line, I realized it couldn't possibly be a bear because it stood taller than the young trees. My friend and I had not heard of any sightings so we didn't have any preconceived ideas of what we were looking at. We continued hiking to the summit, mainly out of stupidity. We didn't talk about what we saw or what it could have been. When we hit the top of the tree line, I realized that there was no way a person could have ran down the face of the mountain the way we witnessed that thing do it. The face of the mountain was covered with loose rocks and gravel, making it hard to descend at a fast pace, let alone run! When we returned home, we talked to our roommates about the sighting and a friend said it must have been a Sasquatch. We returned the next day but didn't see anything. We did, however, take note of the surroundings. The trees at the tree line were well over 7 feet tall, and the creature towered over them. The area in which it ran to is a place where no human can get to. There are no trails, and it is too steep to try to climb down. The forest there is thick and green and anyone who would stray off the trail could easily become lost."

The main witness in this case did at one point have his camera on the subject but did out snap a picture as, at the time, he thought he was looking at a man. He did search all of the other photos he had taken at that rest stop in case he might have caught the subject unintentionally. Unfortunately, this was not the case. Both witness wished to remain confidential.

October, 1985

This incident may have happened in 1985 but I was not made aware of it until the witness contacted me May of 1993. He wished at the time to remain confidential so I will just refer to him as Mr. A.B.

A.B., who resided in Calgary Alberta at the time of the investigation, told me of a strange animal he had encountered while traveling north on the 940 Trunk road, approaching the Livingstone Recreation area, just about 6:00pm, October 1985. He could not recall the exact date. There were two other people in the car with him. His son, a boy of 12 years, and a friend who's first name was Leo. The sun was getting low after a warm sunny day with clear blue skies. As they approached the Recreation area turn off they noticed ahead of them, perhaps two city blocks distance, a strange animal walking away from the east side of the road (car passenger side). They only saw the creature for about 15 seconds as it walked into the lodge pole pine trees. It was not seen again. The two men thought at first that it might be a grizzly bear as they pointed out to Abs son. They thought it strange the bear was walking the whole time it was in sight upright on two legs? A.B. Described the creature as covered in cinnamon colored hair. (Yellowish Brown) and that it appeared to be well over 6 feet tall, He described a flat face, however he also stressed that they really did not get close enough to notice much of the facial detail. He thought it was a very heavy creature but wouldn't hazard a guess as to how heavy it might have been. But he did state, "It was god dam big!" He did stop the car when the spot was reached. They then cautiously searched along the tree line for about 10 minutes for sign as a bear was no longer their prime suspect. No footprints were found and the creature was not seen again.

At the time of my interview with A.B., he was a Commissionaire for the Calgary Regional Heath Authority (C.R.H.A.). Before that, he was a big rig driver from the early 1960s until retirement in the mid-1980s. His fellow commissionaires described him as a friendly fellow who was not known to be a prankster or story teller. In fact, at the time of his report to me, he still was not prepared to state that it was definitely a Sasquatch they saw that day. He can't figure out what it was. Another interesting note to this report: this incident happened almost on the same spot Mr. Adlington claimed to have had a sighting eleven years later on August 4th, 1996, the first incident in this update chapter.

May, 1990

This Alberta report was sent to B.F.R.O. Researcher, Dawn Harrack, who thought I would be interested. She was right, of course. This case was two American young men driving north through Alberta on their way to Alaska, ten years earlier (May 1990):

"We were about 90 miles south of Jasper, close to the Saskatchewan river crossing. The creature was on the right-hand side of the road as we were heading north. My friend, Jason and I were on our way to Alaska at the time from college (he to go home, and I to go to work for the Fisheries). As I said earlier, it was about 45 minutes before dusk. We had just come around a bend in the road when we happened to look to the side of the road and saw the creature. It looked as though it was standing still against a tree in order not to be seen, as it undoubtedly heard us approaching. It looked right at me as we passed it at about 45 - 50 MPH. In the approximately four to five seconds we saw it. I could make out the shape of its face and the color of its fur (Irish setter red); in about 3 strides it turned and was out of sight. I do remember its eyes were yellow, like those of my wife's cat. It did not look friendly, if that is the right word to describe its demeanor. It gave me a look like a 'wild dog' might give if you came upon it feeding. I know it's been 10 years since my sighting but my recollection of it is fresh, as it is one of the most profound experiences of my life."

The witness in the case did not want his name made public at the time of our talk so I assume the same was true for his friend who was actually driving the car at the time. The friend did not make out any report for me or the BFRO. I was most struck by the hair color described: Irish Setter red. I have heard of reddish-color hair reported before but this was the first and only time in my experience that I can recall Irish Setter red described.

November 17, 1995

This incident did not come to my attention until the year 2000. I was visiting the late John Green at his home in Harrison Hot Springs when he told me of this Alberta report he had received from the BFRO. John wanted me to follow up on it. The witness did talk with BFRO researcher, Kevin Withers. There was contact information in the hands of John Green and when I returned to Calgary I contacted the witness in this case, a man named McAvena, who now resided south of Calgary in the town of High River. I paid him a visit in August 2000 to interview him about what he saw on November 17, 1995 while hunting off Route 532, not far from the town of Longview. What follows are parts of that interview. I found Mr. McAvena to be honest man whom was more curious about he saw than I was.

"Tell me in your own words, what happened?"

"We were across the valley, my dad and I on the other side, and there were a bunch of trees up through the bottom and up the other side, and then there's a big open field on the other side. We were sitting there on the little knoll there looking for animals. I had my binoculars and what not, and about a quarter mile away, in the bush, with my binoculars, I saw two mule doe (mule deer, female); I had my mule doe tags. It was the afternoon, it was after lunch, probably around 1:00pm or so, and I spotted these two deer and I said to my dad, there's two deer. He says, where? I had to explain to him where they were, and he finally saw them in the trees there. When they came out of the trees there, he says, oh yeah, then I say, I'm going, and he says, you're not going to make it over there, I say sure I am. But I went up, and it took about, oh, probably 10 to 15 minutes, to walk up there. There's an old logging trail that comes up the side of the hill to the top. I walked along up that road, I wanted to get up behind them (deer), 130 150 yards away from them. There was a lot of snow, the snow was about a foot and a half deep, and I was trudging along. I got up there and I had a shot and I took one. Unfortunately, I missed for whatever reason; probably hit a tree or something, because there were trees between us. So, they went up and around, there's a gorge, and they went up and around the gorge where the open field was. So I thought, well, I will go off after them, figuring I would catch them somewhere in the field. I get there, and I start looking around, went up along the edge of the bush and started to look across the field for these two deer. Of course, I couldn't see them at all. At the far end of the field, probably close to 400 yards away, I spotted these trees swaying back and forth. It was a very calm still day and these trees were swaying back and forth, and I thought, 'what in the world'? I raised my rifle, figuring a bull elk or moose or something, I raised my rifle. I have a 10 power scope on my gun, two and a half to 10. I'd turned it down to two and a half because I was in the bush. I didn't turn it back up, so I was looking through the scope, all of a sudden this creature comes out from these trees, with its hands or arms straight up in the air, swaying them back and forth. I thought, 'here's some idiot out there dressed in black waving his hands scaring the animals away!' He was far enough from me that he didn't concern me, so I put my gun down, because as soon as I saw a human form, you know, you put your gun down; you don't stand there with a loaded gun aiming at somebody. So I put my gun down and looked up and it took maybe, oh, 10 seconds maybe at most, to run from where it came out of the trees, across the field and down into the bush again, a distance about 100 to 150 yards, something around there. I was thinking at the time, it was some idiot, some environmentalist, out there chasing the animals away so we couldn't hunt. There had been reports that some of these people were doing that. So I turned around and started heading back through the bush. Going through the bush I didn't see anything, it was real quiet. I went back to where my dad was, it was about three o'clock in the afternoon, so started heading back towards the truck. We were about a mile into the bush. I told him what I saw. We both figured it was some guy dressed up chasing animals around?"

The two men returned to the same area the following day and both went to the treed area where the trees were seen to be shaking. The trees seemed to be at least seven inches thick and not all that easy to shake. Mr. McAvena then went to check the line of tracks the figure made as it walked away. It was deep snow so nothing resembling the bottom of winter boots was seen. Also, it took him nearly 25 minutes to cover the same snow-covered gorge the subject seemed to cross in a few minutes. It was his father at this point who first suggested "Sasquatch". Mr. McAvena did say it was all black from head to foot but he really could not tell if it was covered in hair or not. I had no reason to doubt that this man saw something or someone that day.

September, 2000

On January 21st 2001, I was contacted by a young man whom did not wish to identified, so I will refer only to his first name, Seth. At the time, Seth lived with his parents in Grand Prairie Alberta. He was 13 years old. His father encouraged him to report what he and a friend of 15 years saw while they were out on two ATVs on Seth's grandfather's land near Rycroft, Alberta the previous September of 2000. The other witness goes by the name John. On the date in question, both boys were riding ATVs not too far from the grandfather's cabin. Both boys were armed; Seth carried a 12 Gauge shotgun while John carried a 22 caliber rifle. They stopped and were discussing where they should go, and in what direction, when they both were startled by a strange high-pitched animal scream, which seemed not too far distance. They started up the ATVs and headed in the direction they thought the noise was coming from. After about 10 minutes, they came across what looked like footprints on the ground. The prints seemed to resemble barefoot human prints but bigger, 15 inches. The boys did not try to photograph the tracks at the time, as they had no camera and the cell phone was of the old flip type, which did not have a built-in camera. They followed the broken line of prints down to a creek when they both saw the subject walking along the opposite bank. Seth thought the subject had not seen them at first as it was walking, and then it would stop, walk a few more steps and stop again. My thoughts are, it must have heard and seen the two ATVs. When it seemed to see them, it turned and dashed into the trees to disappear from view. John told me, "It wasn't moving at all and it just stood there in one spot looking at us for a few moments then turned and slowly walked into the trees". I have to say, a little different than dashing into the trees described by Seth.

Both boys said that the subject was hairy; both say that the hair color was brownish-black. Both boys said that the subject stood and walked upright on two legs, and that at no time did they see it drop down on all fours.
Seth claimed that the encounter lasted one to two minutes; John says two to three minutes. Both boys describe the face as gorilla-like. Seth saying the skin color on the face was very dark. John says the skin color was black.
Neither boy smelled anything before, during, or after the subject was no longer in view.

Now, Seth's father told me after I was finished talking to the boys, that they had contacted me at his encouragement as he felt neither boy was the type to make something up and stick to it; they why they were. But he did mention that Seth's grandfather (who owns the cabin and surrounding land), does like to tell Sasquatch and other strange stories to entertain around the cabin fire at night. Also, John was reading my second book at the time they contacted me (Sasquatch/Bigfoot the Continuing Mystery). I was inclined to believe that the boys had indeed encountered something or someone that day.

September 2, 2010

The Crandel Campground incident of May 1988 remains, for me at least, one of the most interesting and convincing cases I have ever looked into. The original publication of this book in 1989 was delayed for a bit so parts of the witness's interviews could be included. Not too far from Crandel campground, another incident occurred long after I had left Alberta for the Lower Mainland of British Columbia. I remember thinking in those years of research in Alberta how unfortunate it was that myself and the late, Professor Vladimir Markotic, seem to be the only people looking into reports on the east side of the Rocky Mountains. After Vladimir s death, I seemed to have the whole place to myself. Since the move to the west coast, a number of good researchers have taken up the cause. Sean Viala is one of those good researchers who have continued to look into matters, He created the Alberta Sasquatch.com web page, and he informed me he was investigating a roadside report inside Waterton Lakes National Park, very close to the Crandel Campground.

"Thomas

Here are the emails back & forth between myself and the witness.

I briefly spoke with him on the telephone & he seems like a credible indi-vidual. The person that sent him my way is also I consider credible. I have known her for a good part of my life. (I met her when I was 7 years old, I am almost 33 years old now.) I let him know I would be sending you the emails regarding this sighting. I figured that you might be interested, as the Crandel sighting was one of the most important occurrences of your research career.

Sean Viala"

Sean did correspond with the main witness in this case after assuring him that he would not be identified. That assurance will continue now as he will be re-ferred to as MS. Here is his response to Sean's inquiries:

" Hello, my name is MS. S------- told me that you are interested in Sasquatch and gave me your email address. Here is my story. My girlfriend and I de-cided to go camping for the September long weekend. I searched the internet for the best place to go, and decided on Waterton Lakes National Park, as neither of us had ever been there before. We would choose our exact place to set up camp when we had arrived there and looked around for a bit. We left on Thursday afternoon around 1 pm. We did this so we could get there and get things set up before all the other long weekend campers began to roll into the area. We arrived in Waterton around 4 pm and began looking around the place, for a nice spot, it happened as we were driving along Cameron Falls Drive about two kilometers from Crandel Mountain Campground. As we came around a slight bend in the road, my girlfriend and I spotted what we thought was a person that was about to cross the road, so I slowed down just in case they tried to cross in front of me. As we got closer, I realized that

it was not a person. My first thought was that it was a bear standing up on its hind legs. My girlfriend didn't know what it was. Then it crossed the road in front of my car and we knew it was not a person. It only took 3 to 4 steps to get all the way across the road, then it walked off into the trees on the other side. I wanted to stop in the next campground and have a look around but my girlfriend would have none of it, she wanted to camp as far away from where we were as possible. She was really freaking out, so to calm her down I just kept driving. We finally camped at a spot that was away from any campground and I had a good weekend. She was nervous all weekend, every noise she was jumping at. We didn't see anything else that weekend except for the deer and a small bear. If you have any questions, please ask. I will try to help as much as I can, but it was over and done with pretty fast.

Yours truly,

MS"

After several more emails Sean informed me that both witnesses in this case did not wish to be involved in any further investigations. They also wished not to be named publicly. They did give Sean some reasons for this request and I can't blame them.

Last week of May, 2004

Just after hearing of the case you have just read, another roadside sighting report was forward to me which had also occurred inside the boundaries of Waterton Lakes National Park, though it predated the previous incident by six years.

"It was the last week of May 2004 on the Akamina Parkway in Waterton Lakes National Park. Early in the morning, it was overcast, light drizzle. I was driving and remembering thinking to myself that there were a lot of Snowshoe Hare (rabbits), along the side of the road eating the fresh greens. I focused on one Hare on the left side of the park way, but my attention quickly focused on the right side of the road and I saw a large black figure looking at the Hare and then quickly looking at me. It was standing in the ditch with its left arm turned away a branch as if to get a better look at me, then quickly as it happened it ended. I kept driving past wondering to my-

self what I had just seen. I was in a daze of sorts trying to comprehend what I had just saw? I did not stop to try and get a picture, nor did I stop to look for prints. I was just in a wild state of shock. That's all I can say. It's hard to describe my feelings. I have only told a handful of people. Most laugh. I still think of it today, playing the whole scene over in my head. When I returned home from my photography trip, I researched the internet and found there was a sighting at Waterton almost to the day in 2002."

This report was given to me by another good Alberta researcher, a fellow named Rob McNeill. Rob told me he considered this witness as a reliable fellow, though he did not wish his name to be made public. No problem there, he is with the majority. The witness was of the opinion that the creature he saw was hunting the Snowshoe Hare which seemed to out in force along the road at this time of year. He saw it in a bush-covered ditch, from the chest up; black hair, solid black. He estimated that with the ditch being very deep that the creature was about 7 feet tall, though he stressed that this was speculation on his part. Was it a Sasquatch or some other animal he mistook for a Sasquatch? Rob McNeill is of the opinion that he was telling the truth. I would agree with him.

May, 1992

I received a letter from a fellow whom asked that his identity remain confidential. So I will refer to him here as Bob. He made contact with me during the month of December, 2016. I later gave him a phone call, as he told me this was concerning something he had encountered in Cypress Hills Provincial Park, 24 years before in 1992. Well, my thoughts were no hurry here, not like there will be anything to follow up on so I took my time over the next couple of days and prepared a file number for Bob's report (Steenburg File# 50094). But as the details were reviewed, I had the overwhelming feeling that this story was familiar and that I had heard this before. So, I started going back through my Alberta report files and, sure enough I came across Steenburg File #50071, which contained a file report from 1992 that was sent to the late John Green, just after the incident occurred, but the sender would not identify himself. John sent the information to me hoping I could follow up on it. Needless to say, without any contact information, my search went no where and I just ended up filing away what was in the written report. I did not mention to Bob that I had a copy of what he wrote to John Green 24 years before; let's see if the story has altered at all over that time. Bob sent me a detailed written report:

"Hi there, my name is Bob ---- and I live in Medicine Hat, Alberta. My partner and I had a sighting of a Sasquatch in 1992. The sighting happened in the
Cypress Hills at a lake known as Spruce Coulee. The month this occurred was
in May, and I believe it was a Tuesday, and was the last day of the month. The
reason I remember the date was because myself and my partner both worked
the previous three days as it was a long weekend and we both worked.

We had planned a fishing trip to Reservoir Lake which is also in the Cypress
Hills, this fishing trip was planed on the Friday of the long weekend and no
one knew of our plans. We worked our long weekend shifts and could not
wait to get out to Cypress Hills and fish. The long weekend was over and our
day that we planned had arrived, we packed a lunch, got our gear together,
went and got gas, and we were on our way. When we arrived at Reservoir
Lake it was around 10:30 am. We were shocked to see that the RCMP (Royal
Canadian Mounted Police) were on the boat launch watching a very large
boat that was dragging the lake for two men in their early twenties that had
fallen out of their boat and were presumed drowned. The families were also
on the boat launch waiting for word on their loved ones.

It was a very sad sight to see. We decided not to fish while this rescue opera-
tion was going on. We made a decision to go to Spruce Coulee and see if we
could have some fun fishing there, as it was at the time full of stunted yellow
perch. I did not think it was very good idea to go to Spruce Coulee because of
the very large number of very small perch, but my partner came along to fish,
and she wanted to fish at Spruce Coulee. We arrived there at around 11:40
am. When we got there, there was a flatbed work truck parked on the dam,
there was 3 guys working on the bathroom, making a lot of pounding noises,
tearing wood off the walls, anyways, we were there for about 15 minutes
when, the workers stopped their work and threw their tool belts on the back
of their truck and took off in this truck, of what I figured might be their lunch
time. As soon as the truck was out of sight, I happened to be looking at the
path that runs the east side of the lake.

This is when our sighting occurred. I saw it first. My partner was a little ways
from me to my right. She was fishing when this very large two-legged animal
appeared on the east path out of the woods. It was quite a distance when first
seen by myself. I immediately froze, could not deal with what I was seeing,
and all the hair was standing on my body. After I thawed out from fear, worst
fear ever for me, I watched as this animal was heading north towards me and
my partner on the civilian trail, and it was moving very smooth for such a
large animal, and was covering a lot of ground with very large steps.

When this animal started to get a little closer that is when I notified my part-

ner, for her to come and see this unusual thing walking on two legs. When she saw it, it was approximately 300 yards straight across from us. She guessed it to be between 7 and 8 feet tall. This animal was alerted by our talking and had walked another 100 yards towards us. But all of a sudden, it went into a crouch ad started pulling weeds and grass. This animal's hair color was a reddish-brown, not very dark. What ended our sighting was a very thick layer of fog had crept over the lake and we lost sight of our animal, and then we hustled to our vehicle and got the heck out of that area."

Bob ----

I never told Bob that I had a copy of the report he sent to the late John Green. I was impressed how little had changed when I compared the two. He was still bothered by what he saw, but he did return to fish on occasion to Cypress Hills Park. He has never seen anything like it since.

May, 2011

I will conclude this Alberta update with the last case report I have looked into from the province. This incident occurred in a public campground which has had a history of sighting reports, a few of which I did investigate when I was there and am happy to find out incidents are still reported. The witness in this case also requested to remain anonymous, so I will refer to him as AV. I found out about this incident when the witness sent a report to the website of researcher, Cliff Barackman. Cliff, remembering my history of research in this area, forwarded a copy of the report to me. I did try to phone AV but it was some time before he heard my message as he and his family was on vacation in British Columbia for awhile. He did contact me on the night of May 6, 2013, nearly two years after the incident. I made arrangements to interview him the next night (over the phone). I will include the whole interview here to give the reader an example of how I conduct a witness interview and also how the questions have both changed and stayed the same since this work was first published in 1989.

Could you state your full name please?
AV

What is your present mailing address?
Answer withheld.

What is your phone number?

Answer withheld.

What is your occupation AV?
I work for an auto company's parts department (company name withheld).

You are former military, aren't you?
Yes.

What branch of the military?
Army. 2nd Battalion P.P.C.L.I.

Do you wear corrective lenses?
Occasionally, for reading.

Were you wearing them at the time this happened?
Yes.

What was your age at the time of the incident?
I would have been 30.

Please tell me what happened?
Basically, what happened is we went camping. It was the first time we had gone camping that year. We had got to the campground that Saturday morning. We spent the day there, we, you know, did typical camping things: walked the dog, had lunch, we had just finished up with our dinner, when we decided to go for a walk down by the river, we were just kind of walking along the bank for a bit, not paying much attention to anything. The sun had just gone behind the mountains, which was why we decided to start heading back, and I took the dog down to the edge of the river, and when she got down there she just started staring across the river.
I didn't think much of it at this point, told her to get a drink, hurry up, she just kept staring across the river and then all of a sudden the hair on her back started to bristle. I looked over and couldn't see anything out of the ordinary, and she is still staring intently across, so I kind of got on top of her and put my head on top of her head so I could kind of see the direction she was looking at, then I saw this large mass across the river. It was um, 150 or 130 yards away from us. I originally didn't know if it was a large bear or a rock because it wasn't moving. Then after about 30 seconds of looking at it, I bent down to the dog and told her, "Com 'on, let's go", and looked up again and that's when I saw it stand up. When it stood up, it took a step to it's, would have been to it's left, kind of pivoted, and took three more steps into the tree line. It was three distinct steps I saw, and when it went into the tree line, into the shadows, it just blended so well with the shadows of the

trees, I did not see it anymore.

Where did this occur?
Bow Valley Campground, which is Bow Valley Provincial Park, just east of Canmore, Alberta, just off the Trans-Canada Highway. It's right on the Bow River.

Near the old cement plant on the river there, right?
Yep. Just down river from that a little bit.

What was the date of this incident?
We always go camping, um. It was the weekend before the May long week-end (Victoria Day in Canada), I think it would have been May 10 or 11th? The Saturday before the long weekend. That's the first weekend we go camp-ing. (Witness at this point starts going into details which I have not asked for yet). When I first saw it, I, you know, I know what it wasn't, but I didn't know what it was. And I didn't think much of it. You know my thought on the situation was it wasn't because there are none (Sasquatch) in Alberta. That's more BC, Washington, Oregon area, so I didn't make a report for it at the time. It was later on that year I was actually reading through the paper, and came across an article about Todd Standing, and his accounts and everything, and looked at his research and what he had was NOT what I saw whatsoever, so I put the whole thing by the way side for a bit. It was probably close to a year after I saw it when I started talking a little bit about it. I made one report to another group of research people, and we did an interview and went out to the location. He told me (BFRO researcher, Gary Cronin) that there had been other activity in that area that same month reported by other people.

What time of day or night did this occur?
It was just after the sun went down. Between 8 and 9 o'clock. The sun had just gone behind the mountain and that's why we were heading back, so it was still light enough that you could see, but there was no glow from the sun in the west direction.

Describe the area in which this took place?
It was off the Bow River. Fairly large river.

Was it on the south bank, or the north bank?
We were on, we would have been on the east bank. And it would have been on the west bank.

Does not the river run west to east at that point?

Um. Yes it runs west to east.

So was the subject on the north or south bank?
It would have been on? I am trying to think here. Would have been on the north bank.

[I think AV was wrong here: the campground lies on the north bank of the Bow River, which is where he was camping. If the creature was on the opposite side of the river that would have to be the south bank, unless of course, camp sites now had been added to the south bank since I moved from the area in 2002].

-- We continued:

So you are saying it was on the north bank?
Yeah. I think so?

What distance would you estimate you were from the subject when you saw it?
115 to about 130 yards.

What color was it?
Ah, really couldn't get a color, not a specific color, it was dark, but not black. Possibly a heavy dark brown? When I thought it could be a grizzly, but it was darker than a grizzly, wasn't a black bear, way too big for a black bear. Ah, you know with greyhound there's a color which is called Dunn (I did not know that), which is gray with some brown mixed with it, it was similar to that.

What was your first impression of it?
I thought, because it was hunkered down, I first thought it was either a rock or a grizzly bear, ah, but when it stood up, that's when I was 100 percent sure of what it was, I knew what it wasn't. I knew it wasn't a person. Just from the size of it. I know it wasn't a bear. Because a bear couldn't stand on two feet and pivot the way this did, and walk as clearly with a stride.

How did you react when you first saw it?
Shock more or less. Just trying to process, what was that? I just saw this and I can't explain what I saw. I felt no fear or excitement or anything like that, it was more military, you trying to evaluate your situation and surroundings, ah, I just tried to figure out what it was I saw. I really can't explain it.
What was it doing?
It was bent down near the river. I don't know if it was digging at the river,

as soon as I, when I was looking at it there was no movement from it what-soever. Then, as I said, after about 30 seconds, I turned around and went to grab my dog and looked away from it, when I looked back up that's when it got up and moved, so I don't know if it noticed we, were over there, and was just staying still? When I broke looking at it, that's when it decided to get up and move.

Was it hairy?
I really couldn't tell if it was hairy? It was one solid color. At 115 or 130 yards I really couldn't distinguish hair, it wasn't skin, it wasn't clothing, I can rule out what it wasn't, but I can't say with 100 percent certainty that it was hair?

How far did you think you were from it?
115 to 130 yards.

Did it stand and walk upright?
Yes.

Did you ever see it go down on all fours?
No.

How tall do you think it was?
Based on the size, and the distance, ah, 7 and a half? Yeah I would say 7 and a half foot range.

How much do you think it weighed?
It would be 500 pounds. Easily. Shoulder width, um. Being in the military you see guys that distance walk with full rucksacks, and kit gear, with 130 pound kit gear on their back and this thing was bigger than that.

How long did you see it for?
30 seconds when it was standing, no, when it stood up took those 3 steps into the bush it was 8 to 10 seconds?

When it was down by the river, was it at the water's edge?
Right at the water's edge, yes.

How far from the water's edge was the tree line?
Ah. I didn't go over there. I estimate 10 yards?
Did it see you?
I think it did.

What was its reaction?
I think its reaction was to stay as still as possible. Which it did until I broke eye contact with it? As soon as I did that, that's when it got up and moved into the bush.

Could you see any facial features?
No.

Could you tell if it was male or female?
No.

Can you describe the arms?
When it walked it did have arm swing, you could see that, you could see the motion.

Did it make any noise?
No. No noise.

Did you smell anything before, during or after?
I didn't. But I honestly think my dog might have? Which might be why she started looking over there?

What breed is your dog?
She's a greyhound.

How old?
She would have been three and half when this happened.

Did she bark or growl at the subject?
Not at all. Which is strange for her.

Are there any other physical characteristics of the subject that stand out in your memory?
Just the fact when it did get up, it made a step off the left hand, I mean left leg, I mean it made a distinct pivot, 180 degree back toward the tree line.

Did you check for footprints later on?
I did not. No, no, not knowing what it was. My daughter was there but she did not see it.
Was she at the river bank with you and the dog?
She was at the river bank. She was throwing rocks in the water.

You didn't draw her attention to it?

Well, I didn't know what it was?

How old was she at the time?
She was 8.

What do you think it was?
Originally I thought, I didn't, I thought, looks like a Sasquatch or Big-foot, but at the same time I thought they don't live here. They are on the west coast. I always thought that was the reported range or habitat? After awhile, after I processed everything it came to me that more than likely that's what it was.

Have any of your friends or family members ever had an incident like this before?
No.

Are you convinced you saw a Sasquatch?
Right now, the more I process this, yes.

Have you ever made a report like this to anybody else?
No.

Do you wish to remain confidential?
Yes. Just because of my work. The reason I haven't mentioned it to a lot of people is because of my profession. And my family I would like to remain confidential.

AV it seems to me did indeed encounter something strange at the Bow Valley Campground that May evening, something which he later tried to find out more about. Studying videos from a researcher whom I personally have grave misgivings about? An individual who claims many encounters with Sasquatch but I personally do not believe a word of! AV says that the creature he saw was nothing like the videos this man claims to have shot over the last 10 years or so; probably a very good reason for that. Fortunately, he did end up talking to another researcher and BFRO member, Gary Cronin, who did go to the location with AV to do some on-site investigation. When I talked to Gary over the phone about this, he told me in his opinion AV was being honest and forthright. I am a little surprised he did not mention sending in a report to researcher Cliff Barackman of 'Finding Bigfoot Fame' who informed me of this incident in the first place. But remembering back to my days in the area all through the 1980s and 1990s, there were several reports in and around the Bow Valley Campground as well just a little further down the highway in Banff National Park.

Well, a lot has happened in the Province of Alberta since I moved away in September of 2002. When I started research there in the late 1970s, I had the whole place to myself. Now, a good number of dedicated researchers have taken over trying to find an answer to the ongoing Sasquatch question on the east side of the Rocky Mountains. There was never a doubt in my mind, as I researched this wonderful wilderness, that if indeed the Sasquatch did exist and was more than a great piece of western mythology and folklore, it does indeed dwell in Canada's Wild Rose Province.

*Any one wishing to contact Thomas Steenburg is invited to
do so at sasquatch@telus.net
or www.thomassteenburg.com*

Author at Pullman Washington, attending the ISC conference
held at Washington State University in 1989.

www.ingramcontent.com/pod-product-compliance
Lightning Source LLC
Chambersburg PA
CBHW051840020726
47502CB00005B/1882